Birthday gift to Nancy from
Robert & Margreet K—
Jan. 11, 19

Master Builders of the Animal World

THE BEAVER AND HIS FAMOUS LODGES.

FROM AN OLD PRINT, 1755.

Master Builders of the Animal World

by David Hancocks

HARPER & ROW, PUBLISHERS
New York, Evanston, San Francisco, London

for KAY and MATTHEW

Contents

Acknowledgements

It is a sincere pleasure to record my gratitude to all those people who have been of assistance to me in writing this book. In particular I should like to mention: Dr Donald Bruning, Dr Fae Hall, Lloyd F. Kiff, Professor E. Kullmann, Charles Mapleston, John Peacock, John F. Preedy, Dr Joyce Pope, Alexander F. Skutch, Dr V. G. L. van Someren, Michael Tweedie, Dr Peter N. Witt, and the library staff of the Zoological Society of London, all of whom have voluntarily offered their time on my behalf, and given me the benefit of their advice and experience.

The publishers wish to thank the following for the use of their drawings and photographs: Australian News and Information Bureau, 73; Bibliotheque Nationale, 76 (bottom left); Dr M. J. Casimir, 58; Bruce Coleman Ltd, 72 (left), 77 (bottom right), 83; Nicholas E. Collias, 17 (bottom left), 49, 51, 52–3; Dr John H. Crowe, 15 (bottom right); Dr Fae Hall, 66–7; Philip Howse, 92, 94 (left); Professor E. Kullmann, facing 80, 123; Fotosub di Malfini e Solaini, facing 49; Charles Mapleston, 74 (bottom right), 88; John Mayo, 44–5; New York Zoological Society, 132; Laurence Perkins, 62 (bottom right), 69 (bottom), 104; Photo Researchers, Inc., 16, 18, 32, facing 32, facing 33, facing 81, facing 96, 98, 101, 112; John F. Preedy, 74 (top right); Raymond J. Quigley, 23, 25 (left), 135 (left); Alexander F. Skutch, 15 (bottom left), 38 (bottom left), 42 (right), 74 (left), 110 (right), 131; Dr V. G. L. van Someren, 13 (left), 135 (top right); P. O. Swanberg, 12; Michael Tweedie, 20 (left), 29, 38 (right), 62 (left), 72 (right), 84, 113 (top), 117 (left), 124, 125; U.S. Division of Wildlife Services, 77 (bottom left); Richard Waller, facing 48; John Warham, facing 97, 109, 135 (bottom right); Dr Peter N. Witt, 117 (right), 120–1, 122, 126 (right); Zoological Society of London, frontispiece, 89.

Introduction

When I first heard that David Hancocks, an architect, was writing a book dealing mainly with animals rather than architecture, I was a little surprised. On reflection however, I realised that he could have definite advantages over a trained biologist. His approach to the subject of animals that build would be that of an architect with an architect's appreciation of animals' building abilities, their selection of sites, their use of different materials, their means of improvisation and the methods by which they deal with such environmental problems as temperature and humidity control and pollution.

The number of examples of building in the animal kingdom is enormous, ranging from the very simple to the very intricate. The sheer technical virtuosity and structural ingenuity is astounding. A broad survey such as David Hancocks's shows a variety of styles and structures from the beautifully proportioned mobile house of the snail to the sophisticated planning of the jerboas' underground township. Even a survey of the building activities of only one class of animals, such as birds, reveals tremendous variety. The majority of bird species lay their eggs in a previously constructed nest, the building of which may involve complex patterns of behaviour and the use of many different materials. Nest building will involve a series of events including the selection of a site, its preparation, the collection of materials, the carrying of that material to the site and the actual construction of the nest. This sequence of events is very similar to man's building activities except that man tends to inhibit his progress by his passion for committees.

In most simple nests only one type of material is used, but most nests are not simple and many materials are used. Materials vary from spiders' silk to mud and plant fibres and even include the saliva of the swifts which hardens on exposure to air. The construction of the nest involves the integration of a number of stereotyped movements characteristic of that particular species, but also involves some learning. Considering the limitations imposed by the almost exclusive use of the bill, the variety of nest constructions is astounding. They range from a slight hollow in the ground with a few pieces of grass to the typical cup-shaped nest with its insulating lining of most song birds. Birds' nests vary in size – the tiny thimble of the Bee Hummingbird is dwarfed by the vast apparently unorganized but annually built-on nest of the Bald Eagle. One example was said to be 8 feet 6 inches across and 12 feet deep, and weighed 2 tons. Though most nests are single structures, some species like the Quaker Parakeet build communal structures in which there are individual chambers.

Primarily buildings are to protect; to protect the inmates, adults and young, and their possessions, including food, from the elements and from their neighbours be they friend or foe. It is these criteria which *should* influence the construction and siting of a building and in the greater part of the animal kingdom these are the factors from which behavioural patterns associated with building have evolved. In most cases, form follows function. Man is the only animal where the motives of prestige and profit, frequently abetted by subjective prejudices, may overshadow the basic reasons for building to such an extent that the results do not fulfil the functions for which the building was designed. Possibly our planners, designers, architects and property dealers could learn from the rest of the animal world. For though its patterns have obviously inspired man, the animal world has still much to teach in the economy and integrity of structure. At the same time it would be intellectually dishonest to attribute anthropomorphic sentiments to other animals' building activities, just as it would be false to belittle the building techniques used as being based merely on a system of reflexes. Emerson wrote: 'He builded better than he knew; The conscious stone to beauty grew,' and though probably only man builds for his own aesthetic pleasure, he is in the fortunate position of being able to appreciate and learn from the many beautiful and fascinating constructions that other animals make.

P. J. Olney
Curator of Birds, Zoological Society of London

Preface

'The swift builds a four-walled home . . . so that its children can live on a hard floor as if they were in their own houses, with nothing to be afraid of and nobody able to put a foot through the cracks and the cold not able to creep upon the tender young.'

T. H. White: *Book of Beasts*

We have long recognised that there are patterns of ordered harmony throughout the world of nature and we have learned to observe that form follows function. We have also attempted to express these truths in our architecture. But too often we have been sidetracked by the pursuits of profit and prestige.

So it is that our history of architecture has been limited to a crippling preoccupation with those works of the western world which were built to commemorate power and wealth. The historian's classifications have ignored the most wonderful examples of vernacular architecture, dismissing them as accidental successes by untutored builders. In consequence this condescending attitude has completely disregarded any of the building abilities of the other animals on this planet, abilities of which this book is intended as a modest reminder.

Superficially it may seem that man as an architect has far surpassed the ability of any other animal. Certainly it is only man who has the capacity to design structures consciously for a pre-determined purpose. But a closer examination suggests that in some ways we may not yet have realised all the achievements of the animals, and in any case it would be presumptuous to assume that we could not derive some encouragement and even advice from the works of the animal builders. Modern architects for example, faced with the universal problem of a housing shortage so acute that even people existing within a damp dark basement (little removed from the caves of our ancestors) cannot be included in the official lists of the homeless, continue to expend their energies for the most part upon the architecture of pomposity and prestige. They do not seem willing to learn from the lessons of economy which have ruled the survival of all other species.

Rather than stop to examine the works of animals and ask *How?* and *Why?*, designers have only turned to nature for inspiration of pattern and fashion. As a decorative motif architects have enthused over the delicate construction of a shell but have tended to ignore the economy and integrity of its structure. They have not become aware that social orders evolved by other species have found their expression in true architecture. Instead man has ignored much of the natural world about him, choosing only to see what he could measure in terms of his own sense of values.

Where there were coral reefs he saw only rich mines of lime deposits; where there were prairie dog towns he saw only cattle traps; where there were beaver dams, beaver skins; where there were honeycombs, mathematical puzzles. He has substituted the beauty of a hummingbird at its nest, or a dancing bird-of-paradise, for coloured feathers in a millinery shop. He has ignored the simple nesting requirements of the chimpanzees and confined them in barren cages.

It is therefore encouraging that today's infant sciences of ethology and ecology are illuminating some ancient facts, and a new understanding of life. As we learn more about the other animals, and understand more about their abilities and frailties, we will hopefully develop an enlightened attitude. Our curiosity and imagination have ensured that we have emerged ahead of all the mammals in the evolutionary race. But that same winning formula can at the same time be our downfall.

Certainly it is no exaggeration to claim that an increased awareness of the way of life practised by all other animals on this planet can only be to our mutual benefit. Moreover, considering the effects which human architecture has on our personal and social lives, it is appropriate that we should carefully examine the works of the animal architects.

D.M.H.

Seattle
February 1973

CHAPTER ONE

Shelter and Environment

Animals and their environment are not independent factors which can be considered in isolation, for all living organisms not only obtain food and shelter from their environment, they also help to create it. Soil for example is not simply composed of inert mineral particles but is formed from organic matter; the carbon dioxide and oxygen of the atmosphere are the products of living organisms; the ocean beds and even the mountain ranges are the massed deposits of millions of generations of minute lime-producing animals. Nonetheless the varied components of the environment are limiting factors on the development of the animal communities. In the soil, on the mountains and under the oceans animals have had to adapt to specific ecological niches. The seemingly infinite diversity in the types of habitat to be found on this planet is thus reflected in the abundant variety of animal forms, for there seems to be no environment which some form of animal life has not been able to colonise.

The study of extreme environmental conditions has provided basic information to help explain more clearly how animals survive and maintain themselves. These extreme conditions however are only human interpretations. Our concepts of normality are not necessarily acceptable even as tolerable conditions by other species. The arid deserts for example, with their lack of moisture and intense midday heat, have not proved totally unfavourable to animal life and a surprisingly wide variety of animals have evolved methods of survival. A great number of small reptiles, mammals and even a few frogs have adapted to the scant and irregular water supply and tend to avoid the heat by seeking shelter underground. Birds nesting in the desert choose sites shaded from the sun, while those which nest in the open, such as the ostrich, lay eggs which are well-insulated with a thick shell. Many species of desert birds regularly construct their nests against the leeward side of small bushes to avoid the sand and debris carried in the prevailing winds. Extra protection is achieved when the nest is then fortified with

masses of pebbles. The ant-chat sometimes collects well over a hundred small stones at the entrance to its nest, and one of the species of wheatears builds such a mound of pebbles that it almost blocks any access to the nest.

Every element of the desert climate is subject to rapid, irregular and violent change. Desert ants have counteracted the severity of these conditions by developing a culture which affords them some measure of independence. Some collect seeds and practise underground agriculture, while others, in the deserts of Australia and North America, collect plant juices from aphids after short periods of rain and store them in the communal stomach of their repletes.

Life in the polar deserts is as harsh and inhospitable as that of the Kalahari. In the Arctic all life is dedicated to survival against the long cold winters, and some animals, such as the caribou, and almost all the birds, survive by migrating. But those animals such as voles, lemmings and the arctic hare have to escape the cold by excavating deep within the snow. Ground squirrels dig burrows to hibernate in the frozen sandbanks, and the great bears are forced to den up. Even when the birds return with the brief arctic summer they are often already mated in order to avoid any delay in nest building. The breeding cycle has to be completed and the birds ready to leave before the dark winter arrives.

One of the effects of such extreme conditions is that they tend to produce small numbers of species, though these are often represented by large numbers of individuals. In more temperate climates there exists a greater variety of species, while in the tropical forests animal life is abundant, much of it adapted to life in the luxuriant foliage of the trees. Ant-eaters, coatis and monkeys clamber among the branches. Lizards glide from bough to bough. Snails, butterflies, snakes, chattering birds, even leeches all abound in this arboreal stratum. Tropical frogs have taken to the trees, glueing leaves together to

Left Nature is unsentimental. In a recent study 86 per cent of all fledglings in an English wood were killed by marauding animals or weather conditions. Open nests are particularly vulnerable, and this wood warbler is disposing of one of her young killed in a hailstorm.

The 'air-conditioned' nest built by rufous-crowned finch-larks, found in the hot and dry tracts of Kenya. The base of the nest is formed of small pebbles, through which any slight breeze can flow. Built against a mound of soda mud, larger pebbles placed round the exposed perimeter prevent the edges from slipping.

Tree frogs in Africa make their nest on the tips of branches overhanging water to protect their young from heavier predators. The nest is made from foam exuded by the female which the male beats into a froth with his hind legs. This sets like a meringue and protects the eggs from desiccation. When the tadpoles develop they wriggle out of the underside and drop into the water below.

form egg sacs. Termites also have taken their nests up to the high levels and build narrow covered passages as their only connection with the ground.

From the highest leaves down through the different layers of vegetation to the deepest tree roots there exists within the forest a complex series of special habitats which have all been taken over by some particular species of animal life. The floor of the forest teems with life: millions of curious animals seek shelter under fungus moulds and the litter of organic debris. There are subterranean species which sometimes occur in staggering densities. Potworms are found in hundreds of thousands to a square metre, and nematodes exist at a density of 10,000 individuals in a cubic centimetre, or 3,000 million to an acre of topsoil. The soil is obviously a safe and steady environment, for the colembella are practically unchanged from their fossilised ancestors of 400 million years ago.

The success of this sort of animal is chiefly due to their size for they, like most of the insects, are concerned with *microclimates,* the temperature and humidity of which can change dramatically within the distance of a footstep. The curl of a leaf can affect the flow of a miniature stream of cold air, and an insect has only to climb a little distance down a blade of grass to avoid it. Some insects spend their entire lives in the shelter of a flower head or, like the leaf-miners, tunnelling inside the leaf of a plant.

At great depths in the ocean the specialised conditions have produced bizarre forms of life, and in the sandy sediment there exist immense varieties of very small animals, blind and unpigmented, hiding in the interstices of the sand grains, living a life little known to man. Extensive modifications have had to take place in these animals to adapt them to the special conditions at the bottom of the oceans. But *all* animals in the struggle for survival have to come to terms with their environment. These modifications are in part structural and in part behavioural. Stream habitats for example are taken over both by fishes, which have evolved mechanisms for contending with the force of the current, and by beavers, which set out to change the character and environment of the stream by building dams. Similarly corals and sponges change their pattern of growth depending on whether they live in calm or rough water, whereas bees have developed artificial structures where they can avoid the vagaries of climate.

Between these extremes all manner of animals seek or construct some kind of shelter within their environment. Some, such as the hare, spend all day lying low in their form, which is simply a small hollow among the grasses, and oil-birds and bats shelter in the warm darkness of caves. But apart from seeking such natural forms of shelter many animals have developed means of providing artificial shelter for themselves, and especially for their eggs or vulnerable offspring.

Emperor penguins breed in the antarctic winter, and in that frozen continent no building materials are available for nesting. They can only shelter the egg from the ice floor by tucking it on to their feet where it nestles under a warm, feathered pouch. For two months the penguins wait in the cold darkness, huddled together against the winter storms on the barren shoulders of the ice cap. Their nesting behaviour is arduous and simple. In contrast the nest forms of tropical species of birds display a remarkable variety: in the weaverbird family alone there is great diversity between the species. Those species of weaverbirds which build in exposed conditions include a thatched roof and an insulated lining. Other nests have thick walls and a sloping roof to keep out the rains. Many species building in trees have developed systems for keeping out predators by adding thick masses of thorns or building long pendulous entrance tubes. The site of a nest as well as its construction can help to give protection. Solitary weaverbirds nesting in forests will often build camouflaged nests concealed among the leaves, while others which have conspicuous structures built hanging from the branches will snip off all the leaves in the vicinity to create a clearing and thus avoid surprise attack.

Not surprisingly many animals have learned to take advantage of man-made structures for shelter. Bears normally den in secluded places such as caves, thickets or in the hollows of fallen trees, but in Yellowstone National Park they have taken over drainage culverts, and one old male used to take up winter residence on the edge of Old Faithful village, dragging his bedding of grasses and broken branches into the sheltered confines of the culvert.

Birds often show a preference for adopting the shelter of human habitations. There are many recorded instances of blue tits taking over letter boxes and of birds nesting in lampshades, raincoat pockets, various machines and on one occasion inside the engine of an aeroplane at Denham aerodrome. There are also numerous examples of remarkable nesting sites taken over by mice: bottles, flower pots, even loaves of bread and carcasses of meat in cold storage. One unusual example of ingenuity was discovered by a farmer who had set a broody hen to nest in a basket. As the days went by it became apparent that the hen was assuming a particularly dilapidated appearance and her ample tail was becoming more and more broken until finally she wore only a stump. Eventually when the eggs had hatched and the hen was removed it was found that a family of mice had taken up residence under the basket. Their nest was made of the material to hand: chopped straw and sacking for the body of the nest with a warm lining of shredded feathers nibbled from the hen's tail.

Within their particular environment, be it coral reef or suburban garden, animals react to only a very small part of their surroundings. Patterns of behaviour triggered off in a conditioned sequence are released by external stimuli. These can be specific

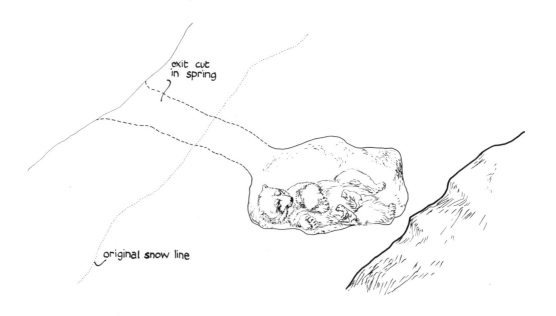

exit cut
in spring

original snow line

Polar bear cubs, born in mid-winter, are blind, helpless and almost naked but are sheltered in the warmth of an insulated den, while the mother waits out her four-month fast until the spring.

Left The nest of the black-cowled oriole is a thickly woven cup of fine fibrous materials. The birds pierce palm leaves and thread strands through the holes so that the nest is suspended. It is then protected from sun and rain by the large leaves and is also out of reach of climbing predators.

Below Many microscopic animals, such as this tardigrade, can survive long periods of drought by adopting a state of suspended animation.

colours or patterns, such as the reaction of robins and sticklebacks to their red-bellied neighbours, while particular calls and songs may elicit other patterns of behaviour. Stimulus can also come from physiological changes within the animal. Canaries for example shed feathers from their breasts at one stage of the breeding cycle, and this naked and sensitive brood patch encourages the bird to add a lining of soft feathers to the nest.

These controlling forces may suggest that the animal acts merely as a conditioned machine and is incapable of conscious expression. Certainly it is true that much of an animal's activity consists simply of stereotyped reflexes and these can persist outside their original function. This is evident in the way a dog pads around in a circle before settling down to sleep on the hearthrug, in imitation of his ancestors who each night padded down a hollow in a clump of spear-bladed grasses on the plains of some neo-lithic settlement. In similar manner young and naked birds make elaborate preening movements and axolotls simulate swimming movements while still in the egg.

These innate actions can easily become confused in artificial conditions. Squirrels in barren zoo cages go through the complex motions of burying imaginary nuts in the concrete floor, while canaries kept in conditions without access to building materials manipulate imaginary grasses and develop other unnatural habits such as plucking out their feathers or repeating stereotyped nest-building actions. Hybrid lovebirds are in a peculiar quandary. Some species of lovebirds cut long strips of nesting material which they tuck into their rump feathers for carriage. Other species use the same materials but carry them in the bill. The hybrids have been observed to try persistently to tuck

Right The European birchleaf roller, *Deporaus betulae*, modifies a leaf to create protection inside a living food supply for its developing larva. After making two S-shaped cuts in the leaf the beetle rolls one half into a spindle with its legs, then wraps the other half over and around it. This elaborate operation takes about an hour, and finally the leaf roller lays its eggs inside the nest, tucking a fold in the tip of the leaf for further protection before leaving.

Below left The most active workers in the animal world are usually found among the birds and insects. The more intelligent mammals tend to be opportunists and it is normally only the small rodents that practise nest building, such as this white-footed mouse which gives birth to helpless and undeveloped young.

Below Nest-building is not merely automatic: it can develop with experience and practice. Compare these two weaverbirds' nests: *top*, built by a young and inexperienced bird, and *below*, a nest constructed by an experienced weaver.

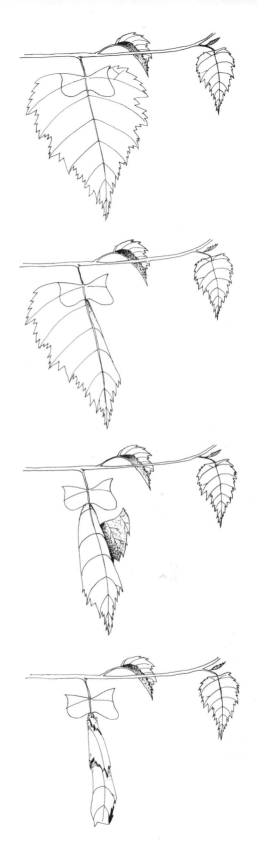

strips of material among their rump feathers but refuse to let go with their bills.

All behaviour displayed by animals is not of course merely automatic. They often clearly exhibit ways in which they have learned from experience. Their apparently simple behaviour can involve an acute awareness of the relationship of things and a calculated response to a given set of conditions. In the sphere of building activities, perhaps more than any other, animals consistently display powers of reasoning and adaptation. Admittedly their capacity to learn is relatively limited, but their ability to adjust their patterns of behaviour to suit changes in their surroundings is not only one of those facets of animal life which commands our respect but, like the chimpanzee which uses a leaf to protect his head from the sun, also represents the true beginnings of architecture: tailoring the environment to suit one's requirements.

The Australian leaf-curling spider making its home in a cocoon of eucalyptus leaves which it has collected from the ground and carried up to its web.

CHAPTER TWO

Mobile Architecture

Apart from any psychological reasons man basically needs clothing to shelter him from the more unpleasant aspects of the climate. This allows him increased mobility and access to harsher environments. Simple temporary structures such as tents and grass huts are extensions of the need to create a comfortable microclimate which is nonetheless transient. Similarly a suit of armour was functionally analogous to a medieval castle, while Westminster Bridge on a wet morning is a roofscape of umbrellas.

From its rudimentary beginnings architecture has evolved through various social forms and in the modern world has developed into the production of substantial buildings for the masses: universities, hospitals, factories and high-rise apartments. But many people in the world continue to live a more simple existence. Their houses, like their clothes, are personal objects. And when they move house they take their buildings with them.

There are obvious advantages with each system, but the popularity of caravan-dwelling and the interest given to plug-in architecture reflects the inherent attraction of a mobile home. Not surprisingly this concept has been investigated along the paths of evolutionary history. The most obvious example of an animal which 'carries its home on its back' is of course the common garden snail whose shell affords protection from accident and attack and at the same time helps to control loss of moisture. There are many other species which benefit from the protection of some kind of personal mobile shelter. Indeed the garden snail is but one of more than 80,000 different species of molluscs, animals which display a great variety of form, size and habit within their group but which share one common feature: they all have a protective, calcareous shell of some type. Shells form a bewildering pattern of shapes and colours. Some are architectonic, or symmetrically formed, while others resemble lumps of stone, gnarled and covered with rough spikes.

The most commonly recognised shell shape is probably that of the scallop. Aphrodite, the goddess of love, was born from such a shell according to legend, and its form has regularly appeared as an architectural motif since classical times. Although adopted as the symbol of St James, the fisherman apostle, scallop shells were also among the unfortunate fashion manias of the last century and were used as ornaments, jewellery and hair decorations. Marcus Samuel was one of the Victorian entrepreneurs who set up a business to meet the new demand. As a sideline he also traded in kerosene and later, when this aspect of his business expanded, the scallop shell was adopted as a symbol for his company. It is now displayed on the oil tankers and service stations of one of the world's largest international oil combines.

Shells are still collected in great numbers by enthusiasts. They are beautiful objects and rare shells can be extremely expensive. When the shells are seen in their natural habitats they are even more desirable to view. Their brilliant colours and patterns are then much brighter. Unfortunately collectors therefore take only living specimens. The animal inside is killed by suffocation or burning and its shell is sold as an empty curio.

Shells are constructed from calcareous deposits secreted by the soft-bodied molluscs, but their shape and form is modified by the environment. Those bivalves which burrow in mud for example have smooth shell surfaces with thick, strong walls. These characteristics counteract the lateral pressures which mud exerts as it accumulates to form a compact and solid mass. Conversely the sandburrowers, having to deal with an unstable, shifting substratum, have developed shells with ridges and furrows for stability. Similarly limpets in exposed areas develop shells which are taller than those of limpets

Mobile architecture at its most refined: the garden snail carrying its home on its back.

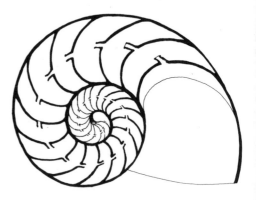

living in sheltered places. This appears to be perverse, for a flattened shell would offer less resistance to the waves, but a limpet living on a rock pounded by the surf has to use its muscular foot more often and more strenuously to keep its place. Consequently the shell has to grow to accommodate the enlarged muscles.

Molluscs range in size from the giant clams of the tropical reefs to the minute planktonic molluscs, such as the sea-butterflies, which swim by continuously flapping their small fleshy wings, or the ianthina snails which float suspended from a frothy raft of bubbles so tough as to resist bursting with a pin. The wide variety of shell shapes are often typified by their names: the elephant tusk shell; wing oysters; the boat-shaped ark shell; basket shells; needle shells; razor shells, and the sunset shell which is marked with faint pink rays. Another unusual tropical shell is the corkscrew, which begins as a spiral but gradually becomes unwound during the snail's life.

Almost all of the more spectacular shells are found in tropical areas, and it is these parts of the world which have the greatest variety of all animal and plant life. In northern climates the lack of variety is often compensated for by a greater number of individuals; the acorn barnacle is a typical example and is commonly found in great numbers on European coastlines. The animal glues itself, usually upside down, to a rock or some solid object and secretes around its body a rigid box made up of six plates locked together with four movable plates on top which act as a trap door. When the tide rises the barnacle opens the doors and with its six pairs of bristly legs paddles for food until the waters recede and the trap door is closed, sealing the animal inside with a bubble of air.

The female paper nautilus, *Argonauta*, grows calcium shells from two of her arms which act as a floating basket for her eggs.

left Cross section through the shell of the chambered nautilus, *above*, shows the pattern of growth of the logarithmic spiral, while the shell of the perspective sundial, an Australian species, *below*, forms the curvature known as Archimedes' spiral.

Classified with the molluscs are the cephalopods, animals of an ancient lineage which were the most abundant forms of life in the primeval seas, swarming in varieties of amazing forms and in thousands of species. They included over 3,000 species of nautiloids which ranged in size from the *plectronoceras* with a shell no more than one centimetre long to the mighty *endoceras* whose shell, the largest the world has known, was at least five metres in length. Today there are only three species of nautilus, confined to the south-west Pacific and the last survivors of the shell-bearing cephalopods. Their beautiful shells grow forward in a spiral, forming a series of ever-increasing chambers which are sealed off with partitions as the animal moves forward to each new chamber. Each discarded compartment is filled with a gas so that the whole structure is semi-buoyant, enabling the animal to swim about more easily. It also has the ability to adjust the pressure of the gas to suit the external forces as it travels deeper. The partitions enclosing each chamber are therefore pierced by a siphuncle running right through the spiral, and the shell is made watertight by the body of the animal which fits tightly into the last chamber.

The shell of the nautilus is often confused with the 'paper nautilus', or argonaut, which has been famous since Grecian times as an intrepid navigator of the seas, sailing in its shell-boat. In fact the shell is built by the animal and used as a receptacle for carrying the eggs, but it is quite easy to imagine the origins of this myth, for the female argonaut has two long arms, expanded at the tops to resemble miniature sails. Special glands on these arms secrete calcium carbonate and the arms are held together as the gelatinous material is slowly secreted to form the delicate egg-nest, paper thin but fluted for strength. Thus the nest resembles a small boat in which the animal seems to be riding. Furthermore the structure is joined on a margin resembling the keel of a boat.

The spiral shell of the nautilus is formed on a logarithmic curve following mathematical laws of growth, and has therefore received special attention from man. There are numerous examples of spiral configurations in nature, from such transitory examples as the flight of an insect around a light bulb or the curl of a chameleon's tail to those forms which actually grow in spiral form: the curved beak of the eagle, the florets of a sunflower and the shell of a snail. The two most important of these curves are the logarithmic spiral, first recognised by Descartes about 1638, and the spiral of Archimedes. These shapes are most easily explained by the diagrams on page 20.

Sir Christopher Wren was among those fascinated by the logarithmic spiral of the snail's shell. He was not the first to discover that the configuration of the shell is a cone coiled around a linear axis, but he did note that the specific form of the shell is dictated by the angle of that axis. There are various relationships between the spiralling curve of the shell and its inclination to the axis, and these determine the overall appearance. In the discoid shells of the nautilus and the well-known ammonites the growth of the spiral is perpendicular to the axis around which it revolves. The shell of the periwinkle however is turbinate, the spiral following a skew path in relation to its axis of revolution.

Because the spiral is a curve whose radius is constantly increasing the shell is able to increase in size with the animal it shelters. It is only the lip of the shell which actually grows, but in spite of its asymmetrical growth the shell does not change its form and so a small nautilus shell is always identical to a larger one.

Other animals than molluscs also benefit from having some sort of naturally occurring protective envelope. Some of the crabs and lobsters have strong and rigid shells. Trunk-fishes and armadillos are encased in armour; crocodiles and rhinoceroses have extremely

tough hides, and insects are protected by an exoskeleton. The unborn young of many species are also protected by some kind of eggshell which takes many forms: birds' eggs are made of a calcareous shell; reptiles lay eggs with tough flexible skins; spiders spin a cocoon of silk around their eggs and shark embryos are protected in attractive leathery purses. The eggs of insects are frequently glued into position and some, such as the mosquitoes', are protected by hard shells which can withstand temperatures below freezing or above boiling point and which can also survive droughts. The female mosquito lays up to 300 of the cigar-shaped eggs at one sitting. These are set upright and are formed into a watertight raft to float on water.

Furthermore animals often lay their eggs in protected places of shelter. Snails and grasshoppers bury their eggs underground. Burying beetles lay their eggs in the body of a dead animal, while ichneumon flies choose the living larvae of other insects. Dung beetles deposit their eggs in balls of manure which they bury in the soil, and some of the solitary wasps construct underground chambers for their eggs and provide them with the body of a paralysed insect or caterpillar.

Sometimes the eggs are also protected by the parents. This is common among birds but surprisingly is also found in such creatures as octopuses, earwigs and mole crickets. Some, such as ants and bees, have the advantage of community shelter, and with most mammals the eggs are protected inside the body of the parent. Other species achieve a similar function by wrapping their bodies around their eggs. This occurs with species as diverse as pythons, butterfish and newts. Female centipedes roll their eggs in sand to disguise them and keep them hidden from the males.

Birds have adopted a variety of methods for protecting their vulnerable eggs. Penguins carry them on their feet, and grebes hide them among floating nests of vegetation. Ovenbirds build enclosing nests as hard as concrete, kingfishers and martins

Right The eggs of the killdeer, like many other
examples which are not laid in concealed places,
have to serve an additional function to protect the
enclosed embryo; the cryptic colouring of the eggs
camouflages them against a drab background.

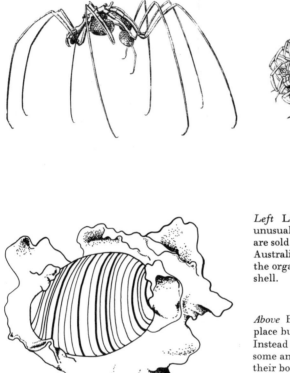

Left Live molluscs are even more dramatic and
unusual in appearance than the empty shells which
are sold as pseudo-scientific curiosities. This is an
Australian marine snail displaying its mantle,
the organ with which it secretes its protective
shell.

Above Eggs are usually deposited in a safe
place but this is not always the case.
Instead of building nests to protect their offspring
some animals offer protection by carrying them on
their bodies. This habit is found in species as
diverse as scorpions and opossums. Here it is
illustrated by spiders, *Nymphon stromi* carrying
its egg masses and *Boreonymphon robustum*
carrying its spiderlings.

burrow deep into sandbanks and woodpeckers chisel holes out of trees. The birds' eggs
vary greatly in shape and colour but never without reason. Those species which lay
their eggs in concealed hollow places have tended to retain the primitive spherical
shape and white colour of reptilian eggs but most others are camouflaged. The casso-
wary for example has emerald-coloured eggs and these are laid on green moss and vegetation. Similarly the pale egg of the ostrich is adapted to its sandy environment. Eggs
laid among pebbles are blotched or dull-coloured, and even blue-coloured eggs, which
are common for many species and which seem glaringly obvious when exhibited on a
museum card, are difficult to see in their natural habitat as their colouring blends with
the quality of light falling through to the shaded nest site. Other special adaptations
include treatments to the shell: thus the eggs of many waterfowl have an oily coating
for waterproofing.

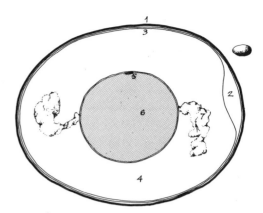

The purity of form of a bird's egg is like the best architecture: each aspect of its construction is specifically related to a functional requirement. The protective shell (1) is porous to allow evaporation of the moist contents which are replaced by oxygen (2), and two membranes (3) further control this percolation. The albumen (4) acts as insulation against sudden temperature changes to help protect the developing embryo (5). As the chick grows it feeds off the yolk (6) and this is kept centred by the chalaza (7).
The egg is drawn to the same size as an ostrich egg, in comparison with a hummingbird's egg (*top right*).

The spherical eggs are the strongest natural shape and also have the smallest ratio of surface area to volume for conserving heat. But only a few families of birds have retained this form. Others have adapted to different environmental conditions. Sea birds, in particular, lay eggs which have an extreme oval shape, allowing them to describe an arc of very small radius should they roll on the narrow cliff edges. Plovers and sandpipers invariably lay four eggs whose shape allows them to fit closely together pointing to the centre. Since these eggs are large in relation to the size of the parent this economical arrangement ensures that the bird is able to cover them all.

Several different factors have influenced the evolution of egg size among birds. The smaller birds in any particular species often have eggs of greater size than their larger relatives. This is because a small chick with its proportionately larger surface of body area loses heat more rapidly than a bigger bird and therefore requires a greater reserve of food before hatching. There are other advantages with large eggs, for although there is a smaller clutch there is a greater store of nourishment in the egg for each developing bird. Moreover the young, like human embryos, are able to spend a greater amount of time in physical development, and in some instances, as with the mallee fowl, are completely independent at birth and able to fly almost immediately. In general the clutch sizes tend to be larger for those birds nesting in holes. The young in open nests are more subject to predation and thus need a more rapid rate of growth to enable them to leave the nest as soon as possible. The adults in any case would be unable to feed a large group of very rapidly developing nestlings.

The shape of birds' eggs can be spherical, conical or elliptical, but the great majority are ovoid, blunt at one end and more pointed at the other. Aristotle claimed that pointed eggs produced male chickens and less pointed eggs contained hens, and this myth has survived for 2,000 years. Early in the nineteenth century Naumann and Bühle put forward the theory that the shape of the egg is directly related to the form of the animal which it contains, whereas Thienemann a few years later claimed that the egg shape is determined by the internal form of the parent bird. Only in recent years has the physical significance of the form of an egg been assessed by relating its mechanical properties to its natural environment.

The shaping of the egg is formed within the oviduct, an elastic tube of muscles along which waves of contraction exert radial forces on the fluid egg. Due to the nature and direction of these pressure waves the greatest forces occur just beyond the middle of the egg. The logical result is that the front end of the egg remains blunt, while the posterior end becomes narrower. The different egg shapes to be found are the result of differences in the size of the oviduct in relation to body size and because some species have stronger muscles. The resultant forms however are always related to ecological requirements. Their specific shapes and the ability of parent birds to produce those shapes have evolved through millions of years of natural selection.

The axial symmetry of an egg is also found in the protective body shell of the sea urchin. This is made up of a number of individual bony plates, all fitting together and growing independently at a steady rate. Although the physical form of the sea urchin's shell is dictated by the same general conditions of forces which determine the shape of a bird's egg, the equilibrium of forces in this instance is due to the constant influence of gravity and not to any external muscular pressures.

The solid body of the sea urchin occupies a very small part of its shell. Most of the space is filled with a dense fluid. It is in a similar state to a drop of liquid in which the form is determined by the forces of surface tension and gravity acting against its internal hydrostatic pressure. Under these conditions a small drop of liquid remains almost spherical but if its volume increases it becomes more flattened. Similarly various shapes of different sea urchin species resemble exactly the conformations of drops of liquid of varying size and viscosity.

In the marine environment of the sea urchin we also find the planktonic radiolarians. These are minute creatures, sometimes large enough to be seen through a hand lens, and their delicately sculptured shells, some of the most exquisite shapes to be found in or out of nature, are architectural forms in themselves. The extraordinary multiplicity of forms to be found in the radiolarians is apparently as endless as the patterns of snowflakes. But whereas snowflakes are symmetrical repetitions of a single crystalline form the radiolarians are each adapted to functional requirements in their environment.

The variety in the symmetrical forms of the radiolarians appears to be identical with the symmetry of forces resulting from surface tensions and it is tempting to consider them as another of those instances in which mechanical forces operate upon a living organism in such a way as to modify it and make it more mechanically efficient. But there are certain species which though still symmetrical display forms which are almost certainly not the result of these tensions. There are regular octahedrons, dodecahedrons and icosahedrons. The reason for these shapes cannot be explained by reference to the principles of space-partitioning such as are manifest in crystallisation. There are no mechanical forces to account for this complex symmetry, and physico-mathematical investigation does not provide any answers. Nature, as Sir D'Arcy Thompson so often commented, keeps some of her secrets longer than others.

CHAPTER THREE

Building by Subtraction

In climates of extreme heat or cold there are distinct benefits to be obtained from living underground. The insulating properties of the surrounding earth help to provide a stable and comfortable microclimate, and in many places throughout the world both people and animals have taken advantage of this. In the Matmata region of the Sahara people live in underground homes. Their administrative offices and mosques are built on the surface but the living quarters and storehouses are hollowed out from excavated courtyards. Similarly in parts of China complete towns are carved out below ground level in the great silt beds.

A great number of animals have adopted the same type of artificial habitat for shelter from climate, and many of these tend to be nocturnal species. Thus the wombat spends its days sleeping at the end of a burrow in an insulated nest made of cork chips. The burrows excavated by the wombats sometimes reach a length of 30 metres. But micro-climatic studies of burrows only half a metre deep excavated by kangaroo-rats in Arizona have revealed significant differences between internal and external conditions. With a ground temperature of 71 degrees C the temperature at the end of the rat's burrow was reduced to 27 degrees C and the relative humidity was generally three to four times greater than the outside air.

The kangaroo-rat stays in its cool and humid burrow during those hours when the conditions above the surface are most unfavourable and subject to greatest fluctuation. In its physiology and behaviour it typifies the way in which animals have been able to adapt to the intense heat of the dry deserts. They tend to be small animals and can move at speed from one shelter to another. They often have long legs to keep their body clear of the hot sand. Typically they are also nocturnal for at night the air temperature in the desert is lowered and there is also a beneficial increase in humidity. Furthermore air movements are reduced, with a subsequent decrease in the rate of evaporation.

To exploit all these features the animals have to seek daytime shelter, either casually under a rock or by digging a burrow. In this respect some animals have evolved more efficient methods than others. The diggings of the jerboas, far from being mere holes in the ground, display sensible planning arrangements of a remarkably subtle nature. The jerboas are small and gentle animals, inhabiting the desert regions of North Africa and closely resembling the kangaroo-rat in form and habits except that their burrows are more sophisticated. They will often for instance plug the entrance to their tunnel with loose sand to avoid detection from predators and seal out the hot air. Sometimes they also include emergency exits constructed on the same principle. Temperature and humidity are controlled by extending the length and depth of the tunnel and in winter they line the nest with shredded wool. To avoid penetration by the winter rains the jerboas dig their burrows in sloping ground, but in the dry summer they move down to the open fields where vegetation is more readily available.

The jerboa digs its burrow quickly and skilfully and uses its whole body as a tool. The soil is brushed away in co-ordinated and rapid movements of its hands and powerful hind legs, and it uses its nose and head as a ram to stabilise the walls of the tunnel. Jerboas also excavate communal burrows where they gather at night for play, but during the day they tend to be solitary.

The gerbil of western Africa is closely related to the jerboa. It also is nocturnal and lives in burrows. But the gerbils live in colonies with extensive tunnelling systems linked by well-defined pathways. The area covered by any particular system varies quite considerably, as do the number of entrances and the depth of the burrows. They

Armadillos, with their powerful legs and long claws, are adapted for digging, and by this means find both their food and shelter. In captivity they are invariably kept on naked floors of reinforced concrete or steel to prevent their 'escape', whereas the provision of a good depth of compact soil would allow the animals to satisfy their urge for digging and would destroy the myth that these animals are inveterate escapologists. In South America other legends persist that armadillos dig into graves for food, a notion which easily captured the imagination of the Victorians.

Far right The larvae of many species of small moths live between the upper and lower layers of leaves. The development of this leaf miner, the larva of a minute moth *Nepticula aurella*, is clearly shown by the tunnel it has carved inside a blackberry leaf. A dark line of excrement remains as a central trail in the mine.

are all however fairly complicated and the presence of rocks causes the layout to become especially intricate.

Gerbils also excavate small chambers which are used for storing food. This habit is shared by many other animals, especially small rodents. The small, brown meadow vole spends most of its lifetime in its elaborate tunnel systems where it stores its collections of food, and many species of ants have evolved elaborate methods of storing food underground. Similarly the burrowing activities of dogs with their bones and squirrels with their nuts are well-known. The concept of using underground shelter for storage of valuable goods has of course long been used by men. Because building in Europe is principally concerned with excluding rain, basements, which are difficult structures to waterproof, have not been developed. Consequently we tend to regard troglodytism as indicative of a primitive culture, but we bury our treasured possessions in underground vaults and make provision for subterranean shelters against bombing.

In similar manner many species of animals in a variety of habitats protect their eggs and young in nests below ground. This is common among a large number of birds. The bee-eaters of Australia drill nest tunnels into the soil, and jackass penguins nest safely in burrows. Puffins and storm petrels are among those species which seem to prefer to appropriate a ready-made burrow, though able to dig their own, whereas the often heavily-scarred beak of the common kingfisher is evidence that it tunnels its own nest, not only in clay banks but also in hard sand and gravel. The common kingfisher's horizontal nesting tunnel is often at least a metre in length, and at its face the bird excavates a nest cavity where it regurgitates half-digested fish bones as a carpet for its eggs.

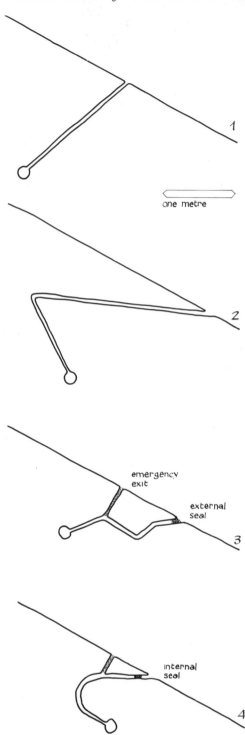

one metre

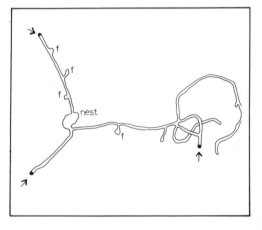

The pileated woodpecker carves its nest out of the stump of a decaying tree. Other species of woodpeckers, however, excavate nest holes in cactus stumps, sandbanks or even termites' nests.

The nocturnal gerbils of south-west Africa congregate in colonies and excavate complicated burrows. This layout shows a typical system dug out of sandy soil. Blind alleys are commonly included, as are small chambers (f) used for the storage of food. Inside the temperature is lower and the humidity higher than the outside air.

Above Studies of jerboas' burrows by J. P. Kirmiz in the western Egyptian desert revealed four distinct types of excavated shelter which the animals create for themselves as protection against both predators and the intense midday sun.

Not all kingfishers however are fish eaters. In fact, of the more than 80 species, most do not live near water. But they all nest in holes; either in river banks, as with the common kingfisher, or in trees. For this purpose they have long, tough and pointed bills, as do the woodpeckers who use their bill as a hammer for splitting bark, as a chisel to cut it and as a shovel to throw away the broken chips. The nest hole excavated by the woodpecker is horizontal at its entrance then drops into a short vertical tunnel. Typically this nest is tunnelled into the branch of a dying tree or rotting stump. In the Arizona-Sonora desert however woodpeckers carve their nest holes in the giant saguaro cacti which are the only plants in that region resembling trees. Although the birds do not cause any significant damage the giant saguaro is nevertheless likely to disappear because of the ecological imbalance caused by the cattlemen who have virtually wiped out the coyote and thus caused a population explosion of small rodents which are causing great damage to the cactus. When the giant saguaro has become extinct the desert woodpeckers no doubt will also die out.

The term 'woodpecker' is something of a misnomer. Apart from nesting in cacti some species are also found in the treeless regions of Africa and South America. These species carve nest holes in sandbanks and sometimes, strangely enough, in the hard walls of termite nests. This remarkable association is also found in many other species of birds, including the jacamars. These are also burrowing birds but their long and slender bills appear to be more adapted for weaving than digging in the earth. Indeed the jacamars are unusual birds in many respects for their bills also appear eminently unsuitable for catching flying insects, their exclusive diet. Another curious but attractive feature of the jacamars is that their nestlings, safely hidden inside their dark burrows, spend much of their time in singing. Similarly young tree-creepers who also benefit from the security of a deeply tunnelled nest often sing loud and harmonious duets in the darkness.

The protection and seclusion of an underground nest is certainly effective. The subterranean nests of some of the nunbirds of tropical America have only been discovered during the past decade. The nunbirds live amid the undergrowth of ferns and shrubs and here they excavate shallow, sloping tunnels into the ground, often arranging a collar of twigs and dead leaves further to conceal the entrance. Unlike the jacamars and tree-creepers however the young nunbirds remain quietly hidden at the end of their tunnel, only coming up to the doorway to take food when the mother bird softly calls them. After being fed the chicks have to toddle down to their nest backwards.

Birds nesting in holes are much safer from predators than those in open nests. Studies of blackbirds breeding in Wytham Woods near Oxford in 1958 revealed a natural mortality rate of 86 per cent of nests destroyed. For this reason birds nesting in open constructions at ground level tend to have short periods of incubation and nestling development, whereas birds nesting in holes can afford a longer time for development. Birds are not the only animals which have learned to take advantage of subterranean shelter, nor are they the only ones to care for their young while they are still under the ground. Some newts for instance make their nests in burrows and here the female stays with her eggs to ensure that they are kept moist. Even some of the dung beetles exhibit parental care. They lay their eggs in deep and slanting tunnels dug beneath horse or cow dung and remain in this claustrophobic cul-de-sac to watch over the grubs until they are safely hatched.

The scavenging burying beetles too are unusual insects in the sense that they provide for their young. They are only small creatures, no more than a centimetre long, but

The energetic scarab gathers balls of dung in
which it lays its eggs, thereby providing its young
not only with shelter but also with a future
supply of food.

The gregarious and colourful bee-eaters
congregate in spectacular numbers, sometimes in
thousands. With the West-African species shown
here, *Merops nubicus*, each pair of birds
co-operates in nest building by digging a
horizontal tunnel several feet in length with an
unlined nesting chamber at the end.

Like the megapodes of New Guinea and
Australia, the alligators practise artificial
incubation. The female constructs a nest of mud
which includes heaps of vegetable material.
The heat given off by the decaying vegetation
hatches the eggs while the mother remains close
by guarding the nest.

they seek out relatively large dead animals, such as a mouse or bird, and working in groups proceed to bury it. If the surface is impenetrable they wriggle their way underneath the corpse and by lying on their backs manage to push it away with their rear legs until it reaches a more suitable burial site. They then burrow under the body. scraping at the earth and shovelling it away so that slowly the cadaver sinks into the ground. The beetles are patient in their work and cut through any obstructing roots, even sometimes gnawing off a limb of the carcass if it is held fast. When the animal is at last well-buried the beetles excavate short tunnels in the walls of the grave. The females lay their eggs in the tunnels and furnish them with balls of well-masticated carrion. Here they wait for almost a week until the grubs are hatched and then help to feed their young with regurgitated food. After two more weeks the grubs, well-fattened, burrow their way into the buried carcass and there they pupate.

Underground nests are also common among mammals, and one of the more elaborate is excavated by the most primitive and curious mammal of all, the duck-billed platypus. The remarkable feature of its burrow is that no soil is ejected. It excavates in river banks, following the line of most favourable soil, and packs the soft soil into the sides of the tunnel forming an arched roof with a flat floor. In this way its winding burrow can reach a length of over 30 metres, although the average length is between five and seven metres. The narrow tunnel entrance is situated just above water level and the sinuous course of the excavation usually progresses about half a metre below ground level, though the platypus will dig deeper to avoid other animals' burrows. At the end of the tunnel the nesting chamber is completely lined with moist eucalyptus leaves and grasses to keep the eggs from drying out. When the nest has been made the female excavates small chambers in the side of the tunnel at various points and with the loose soil constructs a number of plugs. As many as nine plugs have been found in one burrow, although it is not clear what function they perform. They may be for security and perhaps also to maintain humidity within the nest. They vary in thickness from 10 to 30 centimetres, and whenever the female leaves the nest she always remakes the plugs on her return.

The animals which excavate shelters underground are often useful soil animals. Wood ants have legal protection in Germany because of the important role they play in improving the soil, and the burrowing activities of the earthworms mix and aerate the soil, breaking down the organic matter of the land and preventing it from becoming sterile. Earthworms, though obviously terrestrial animals, can in one sense be considered as semi-aquatic. Dry air is fatal to them and they regularly seek out puddles and damp places. They escape from summer droughts or winter frosts by burrowing deep into the ground, but in more favourable conditions they spend most of their day just below the surface, emerging only at night to feed; worms found wandering on the surface during the daytime are usually infected with the larvae of a parasitic fly. However, their habit of lying just inside the neck of their burrow leads to their destruction in great numbers by birds which are able to reach them with their bills.

Worms drag leaves and other vegetable debris into their burrows for feeding but also, and especially during the early winter, gather bits of paper, wool and feathers for plugging the mouths of the burrows. Leaves are also collected in heaps over burrow entrances to form a crude roof. Interstitial spaces between the plug of leaves are filled with moist earth to make the seal more effective. If leaves or sticks are not obtainable the earthworm gathers heaps of small pebbles around the entrance, sometimes plugging the hole with stones of surprising weight. What advantage they gain from this is not

A wide variety of animals, ranging from insects to small mammals, provide underground shelter for their offspring. Here the practice is being carried out by the edible snail.

Plan of a duck-billed platypus' burrow, opened up on the McDonald River in 1920. The platypus excavates its tunnels in river banks well reinforced with tree roots and follows the line of most favourable soil. The entrance, about 12·7 cm in diameter, is above water level and usually screened with reed fringes. The tunnel has an arched roof, with a flat floor, and is usually about 30 to 45 cm below ground level. The tunnel is extended and new nesting chambers are added each year.

clear. It may be done for protection against flooding or to conceal the worm from predators; perhaps also to prevent evaporation of moisture.

The burrow is excavated by two methods: either the worms push the earth around them, thrusting their cylindrical body through the soil as a wedge; or they swallow the earth in front of them. This latter burrowing action is the only way they can excavate undisturbed and compact ground. They also swallow soil as food, ingesting its nutritive value and ejecting the faeces as casts. In this way many of the worms are almost as skilled at tower-building as they are at tunnelling. Some towers are built up to considerable size and may even serve some purpose in protecting the burrow entrance. The tower casts are pierced with a small cylindrical passage which the worm uses as an exit to eject more soil, but this does not run in an exact line with the burrow and the change in direction offers a further degree of protection. Though spending most of their lives near the surface worms are able to burrow to considerable depths to escape from unpleasant conditions. In northern European winters they have been known to travel underground to a depth of three metres. An earthworm's burrow tends to run almost perpendicular from the surface, and is invariably plastered with a thick layer of fine particles of earth which become smooth and shaped to the exact width of the worm's body as it travels up and down its burrow. This lining also appears to strengthen the walls so that a worm's burrow, rather than being a mere hole in the ground, can be considered as a tunnel with a semi-structural lining. The neck of the burrow is lined with fragments of leaves bonded with earth and castings to form a cylindrical case which is about the length of the worm itself.

The earth which is brought up to the surface by the worms is spread out by wind and rain. Gradually these small deposits accumulate and over a period of years, if worms are left undisturbed, their persistent workings cover everything in the locality with a layer of vegetable mould. In this way the earthworms effectively bury all rubbish and other materials deposited on the ground. Darwin recounted how a path of small flag-stones laid across the lawn of his home in Kent evenly and steadily sank into the ground until after a period of 34 years they were completely covered with a one-inch layer of soil by the actions of the earthworms. The specific gravity of the objects does not affect the rate of sinking, and the worms are also responsible for burying very large stones and even ruined buildings in time. In this way of course they have helped to preserve many antiquities. They seem to prefer to burrow beneath solid objects, possibly for protection, and eject their castings around the edges. The heavy object sinks a little each time the worms' burrows collapse beneath them. As the animals continue to expel earth the sinking process continues, which possibly suggests that ordnance survey benchstones may in time become inaccurate.

In many parts of England worms deposit on each acre of land over 10,000 kilo-grammes of earth brought up from beneath the surface. In this way the ground is continually being sifted, loosened and exposed to the air. Bones, shells and vegetable matter are all buried by the earthworms' actions and decay underground among the roots of the plants. Furthermore the leaves which the worms drag into their burrows for food are only partially digested and mixed with the earth form the rich humus which covers so much of the land areas.

It is impossible to measure the vast number of worms living beneath our feet, though Hansen, following thorough investigations, calculated that they live in gardens at a density of 133,000 to the hectare or more than a dozen per square yard. This number gives some indication of the amount of earth they shift and it is doubtful whether many other animals have played so important a role in the development of the land. Nonetheless, because they deface the sterile tranquillity of suburban lawns they are attacked with a frightening array of chemical and toxic weapons.

Moles too are commonly considered as invaders, especially by farmers and estate owners, and systematic attempts to exterminate them have long been waged. Perhaps the only people who ever had kind regard for them were the Jacobites. Their toast to 'the little gentleman in black velvet' was in remembrance of that day in 1702 when the horse bearing William III stumbled into a molehill, causing the fatal fall of the monarch.

The molehill is the only visible sign of the mole's prodigious tunnelling activities and is a common feature of the countryside. But few people have ever seen a live mole, though they are widespread throughout many parts of Europe and Asia and are to be

Cross sections through a mole fortress.
a   A sophisticated arrangement matching the elaborate descriptions which persisted in nineteenth-century natural history accounts.

b, c   Two examples from field studies showing widely different structures, with one fortress containing three nesting chambers. The tunnels average about 5·1 cm in diameter, and the nest chambers are about 20·3 cm diameter.

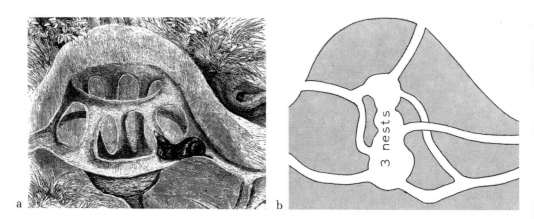

a   b

found in a variety of habitats. In some regions they are numerous and in grassland areas average a density of three or four to the acre.

More than any other mammal the mole is perfectly adapted in all respects to burrowing underground. With its short and powerful front legs it literally swims through the earth with breast-stroke actions. They are active and aggressive animals, cannibalistically unsociable except for the rutting season when the males and females observe a brief truce. They spend almost all their solitary adult life underground, parading the length of their galleries at a steady trot in search of earthworms and attacking any other animal they may encounter. Sometimes they collect live stores of earthworms, biting their fore-ends to cripple them; one inspection of a nest revealed a cache of 1,280 worms. Young moles however spend much of their early life above ground foraging among grass roots. Their first attempts at tunnelling are haphazard and only superficial excavations are made. Eventually they spend more and more of their time underground, safe from predatory birds, and finally they begin their true hunting runs.

The burrowing system created by a mole can become quite an elaborate arrangement with tunnels at different levels being used for different seasons. During the summer when most of the available prey is in the upper layers of the soil the moles make only surface runs. In winter they are forced to dig deeper, burrowing in the soil below the frost line. Some of the runs, especially where there are rich pickings, are permanently in use during certain times of the year and become smooth, hard tunnels. Indeed they may be used for many generations and sometimes a fortunate young mole may discover an unoccupied system of burrows and need never dig its own tunnels.

In the construction of the permanent runs a considerable amount of soil is produced. This loose soil has to be excavated and for the purpose the mole digs vertical shafts from

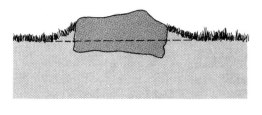

one metre

Above Section showing the extent to which one of the fallen stones at Stonehenge had sunk into the ground because of centuries of activity by earthworms.

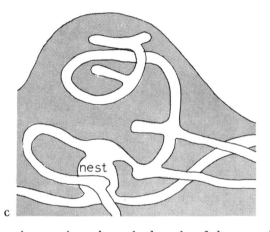

nest

c

time to time along the lengths of the tunnel. It then pushes the soil up to the surface with its broad hands and thus creates the familiar and characteristic molehills which betray the presence of the run. When the mole returns to its winter runs it often has to carry out repair works. The burrows collapse in places because of traffic overhead and the mole may have to excavate a by-pass. If the soil is frozen and unworkable then existing shafts have to be used to get rid of the fresh soil. This is why new molehills sometimes occur even when there is snow on the ground. As an alternative, and especially in clay, the mole may repair a breach by scooping damp earth from the floor and forcing it tightly against the roof, compacting the section into an arch.

As well as operating in shallow runs and deep tunnels moles on occasion are seen running along shallow open trenches. These excavations have often been described as *traces d'amour* though they seem in no way to be connected with the mole's activities during the rutting season. They probably only occur when the mole is digging for food near the surface in unstable soil which collapses to form a ditch. Confusing reports have also been made regarding the mole's 'fortresses'. These certainly exist, though very rarely, and in reality never match the elaborate descriptions given by so many early writers. The following account of a mole fortress is dated 1866:

'The central apartment, or keep . . . is a nearly spherical chamber . . . situated at a considerable depth from the apex of the heap. Around the keep are driven two circular passages, or galleries, one just level with the ceiling and the other at some height above . . . Five short descending passages connect the galleries with each other, but the only entrance into the keep is from the upper gallery, out of which three passages lead into the ceiling of the keep. It will be seen, therefore, that when a Mole enters the home from one of his tunnels, he has first to get into the lower gallery, to ascend thence to the upper gallery, and so descend into the keep.'

A few years later, in 1874, Thomas Bell described the fortress as 'a complication of architecture which may well rival the more celebrated erections of the Beaver'.

In fact the 'fortress' heap, which is quite a large mound of soil, contains one or more nesting cavities and radiating from this are a random number of bolt-holes and hunting runs. Normally the sleeping nest is made in a chamber dug out of the side of the hunting tunnel and lined with dry grasses and leaves. But occasionally, and for some unknown reason, the mole makes its nest inside a fortress heap. Generally these seem to be built by females as a breeding nest. There is also evidence to suggest that these heaps tend to be built more often in areas subject to flooding. However, both the mole and its

The nunbirds make their nest at the end of a sloping tunnel which they dig out of the ground (*left*). Around the entrance they arrange a collar of small twigs and grasses, and a similar arrangement is adopted by one of the spiders of the genus *Lycosa* (*below left*).

The nest of a solitary-bee, *Osmia rufa*, made inside a piece of bamboo. The cells are divided by dry partitions and each cell contains one egg with a store of pollen mixed with honey.

fortress require much more investigation if we are fully to understand its behaviour.

As an animated tunnelling machine there is no other mammal to match the mole, not even the echidnas or armadillos which seem to be able to sink straight down into the earth. But in the insect world there are many other animals equally capable, especially the larvae of numerous insects such as weevils, glow worms, many kinds of beetles, moths and butterflies. The structure of the mole-cricket is completely modified for tunnelling; with its toothed and flattened forelegs it ploughs through the soil in a very similar manner to the mole. Gilbert White some 200 years ago described a mole-cricket nest which had been exposed by a gardener:

'There were many caverns and winding passages leading to a kind of chamber, neatly smoothed and rounded, and about the size of a snuff box. Within this secret nursery were deposited near a hundred eggs of a dirty yellow colour, and enveloped in a tough skin . . . just under a little heap of fresh-mowed mould, like that which is raised by ants.'

White also wrote of the destructive effect of the mole-crickets upon the flowers and lawns of his garden at Selborne, for while they are tunnelling the insects cut through any small roots in their path.

Many of the mining bees also attract disfavour because of their effects on garden lawns. These small solitary bees dig tunnels in the soil to deposit their eggs. This work is carried out by the female. She pulls the grains of earth backwards with her feet and then brushes the spoil out of the entrance hole with her back legs. The tunnel is sometimes surprisingly deep, reaching the length of a man's arm, and at the bottom the bee makes a papery cell and furnishes this with a globule of honey and pollen. The egg is laid on this rich bed and the bee closes off the section. She then constructs a new cell on top of this and continues to repeat the process until the length of the burrow is sealed off in up to half a dozen portions, each containing an egg and its provisions. Her work completed, the female mining bee finally emerges at the surface and there she dies.

There are about 250 different species of bees to be found in the British Isles; only a few of these are social insects building a honeycombed nest. Most of them, like the hundred or so species of mining bees, are small, solitary creatures and display a wide variety of nesting techniques, though many of them follow a similar principle to the mining bees. Some of the potter bees for example also dig in the ground, but at the end of their tunnel they build very small clay pots which they fill with honey and pollen. The egg is then deposited inside and the pot sealed over.

A large number of burrowing animals are adapted for boring in wood rather than burrowing in the ground, and this applies to some of the solitary bees. The carpenter bees make their nests in decaying timber, tunnelling parallel with the grain and using the sawdust from their excavations to make the egg cells. Leaf-cutter bees however prefer to use existing holes, and their nest cells, containing honey and the egg, are made of small sections of rose leaves, which the bee cuts from the living leaf in semi-circular portions, rolled into cigar-shaped tubes. This honey-tight cylinder is sealed off in sections with pieces of leaf, each section comprising one egg cell. When the nest is full and completed the parent bee leaves the tunnel and soon dies. The young bees have to eat their way out and emerge on the basis of 'last in, first out'. The first out in fact are always male bees which, instead of immediately flying away from the nest, merely sit and wait on nearby flowers. When their sisters at last appear at the tunnel entrance it becomes clear why the males have been hanging around.

Many other insects and small animals obtain shelter for themselves or their young by boring into timber. The very large family of beetles also includes a great number of species which are wood-borers. Some, such as the death-watch beetle, are notorious for the damage they cause in seasoned timber, and many of the beetles are pests. The flour beetle for instance gets its name from its habit of burrowing in flour and has been found baked inside bread. Many other species cause injury to stored foodstuffs, cultivated plants and a variety of materials.

The greatest number of beetles, and there are more than 250,000 known species, seek shelter underneath flat stones or bark, and although few people will have ever seen any bark beetles most would instantly recognise the pattern of their tunnel system on the underside of pieces of bark. These can be found on various species of trees, and each type of bark beetle not only restricts its activities to specific trees, such as oak or ash, but each also has its own characteristic tunnelling pattern. The most common is *Scolytus destructor*, a weevil which unfortunately is a scourge of elm trees. Its close relative, the elm-bark beetle, spreads the fungus of the Dutch elm disease.

The tunnel patterns of the bark beetles are produced partly by the adults and partly by their larvae. The sexes mate in a small hollow carved underneath the bark and the female then cuts a trench along the underside. At intervals she lays an egg and later, when these hatch, the larvae tunnel away from the trench, feeding off the bark as they progress. Their arrangement of tunnelling away at right angles to the main trench ensures that the greatest amount of food can be consumed over the smallest area without boring into any neighbouring tunnels.

The tunnelling habits of many beetles and their larvae can cause a great amount of damage. In the home the most serious and common beetle pest is probably the furniture-beetle which if not checked will reduce the interior of pieces of furniture to powder, leaving only a fragile shell. In the tropics the tunnelling activities of some of the termites are even more feared. Many termites cultivate gardens of fungus growths in their nests, and these are prepared on a compost of partially digested timber. To obtain this wood the termites have to leave the nest, but they protect themselves from enemies and from the light by constructing small covered passageways of cemented earth and saliva. In their manner of collecting wood the termites prove themselves to be among the most effective demolition agents as well as being some of the most talented builders. They attack timber structures from below ground and will eventually devour and remove every morsel of wood except for a thin shell, which deceives the observer until the building collapses into dust.

The tunnelling effects of ship worms can be equally devastating. Early wooden ships had to be clad with copper to preserve them from attack but even so many have been lost at sea because of ship worms. The temple of Serapis at Naples, built on a promontory, collapsed into the sea after ship worms had attacked the structural members underwater. Similarly much damage was caused to many of the dykes in Holland during the eighteenth century.

It seems remarkable that ship worms should have developed, in the sea, as wood tunnellers. Before men started to place timber structures in the sea their tunnelling activities would have been restricted entirely to driftwood. The fact that ship worms are so perfectly adapted to their way of life is evidence of the great amount of timber which must find its way into the sea by natural means. Similarly all the other examples mentioned here illustrate how animals found many and ingenious solutions to the problems of seeking a secure shelter, using teeth, nails and beaks to carve out a home, countless millenia before man had to learn how to dig for victory.

CHAPTER FOUR

Building by Addition

It is impossible to make any generalisations about the nesting behaviour of birds or the structures they build. There is such an enormous variety of species living in such a diversity of habitats that they have evolved many different and complex solutions to the same basic problem. Furthermore a great deal of our knowledge has been based on studies made from museum nests, and these do not reveal the variations which may arise within just one species from specific ecological pressures. The class of scientific literature dealing with birds' nests is particularly unsatisfactory and zoologists have restricted their attention in the main to the problems of classification and anatomical structure. It is only in recent years that scientists have returned to the type of studies made by the early amateur natural historians, and are investigating animals and their behaviour in relation to their natural environment.

The earliest known bird, *Archaeopteryx*, appeared some 130 million years ago and evolved from reptilian stock. It had a toothed bill, and claws at the bend of its wing. But it was nonetheless a bird because it had feathers. In fact this is the only general characteristic which can truthfully determine the birds from all other animals. Every other feature they possess is duplicated in some other form of animal life. And some of their seemingly distinctive ways of life are not shared by all bird species.

Not all birds for instance make a nest. The blue-footed booby birds of the Galapagos Islands, like many other species, simply lay their eggs on the bare ground. They do however carry out symbolic nest-building activities. The male parades solemnly around his territory picking up small pebbles and fragments of guano and carries these to the site where he and the female enact an elaborate nest-building ceremony. No structure is built but the birds receive mutual stimulation from these actions which helps to establish a pair bond and ensures that they stay together to complete the duties of incubation and chick-rearing. The different processes of pair bonding between male and female birds cover a variety of relationships, including monogamy, polygamy and polyandry, and all these in some way are related to the many different types of nesting behaviour which are established.

The origins of nest-building remain obscure, but an evolutionary clue to the shaping and building of a nest might be found both in activities of play and in the behaviour and movements of the bird during copulation, such as frenzied pulling at strips of vegetation or making scrapes in the soil. Certainly during the early days of the reproductive cycle the birds seem only to toy and fidget with building material and do not begin any effective attempts at construction until after copulation has taken place. Sexual activity and nest-building activity thus develop along parallel lines, but both before and after nest construction pseudo nest-building activities are often carried out. Initially these may play an important part in the courtship ritual and later act as an outlet for emotion.

Although nest-building is an instinctive ability there is considerable adaptability in both site selection and use of materials, especially with those species which build quite elaborate constructions. Furthermore some element of learning is often evident and juvenile birds do not build as well as their practised elders. Young ravens for example first attempt to build with sticks of quite unsuitable size, while a jackdaw's first nest includes virtually any movable object. John Steinbeck recorded the contents of an osprey nest built in his garden, which included three shirts, a bath towel, one arrow and his own garden rake.

Birds also display remarkable behaviour in collecting building material. Crows have

Above The early morning community-singing linnets, once the caged pets of many a working class Victorian, are now rare in Britain. Their cup-shaped nest typifies the popular concept of a bird's nest: it is built by the female of moss, twigs and roots, lined with feathers, wool and thistledown.

been seen to tear off stout green twigs, and sparrowhawks will dive purposefully on to a branch until it snaps and then hang upside down to break it off. Golden eagles, over generations of work, construct enormous nests. One of these, examined after it had been dislodged by high winds, weighed almost two tons and included foundation branches almost two metres long. The carrying capacity of the eagles however is only relative to their size and most birds are able to carry an extra load of just over 20 per cent of their body weight.

The popular conception of a bird's nest is a neatly constructed shallow cup with a cushion of soft materials lining the interior. This general form is adopted by most of the passerine or perching birds, which comprise about three-fifths of the world's total species. Within the passerines' order there is the greatest complexity and variation on this simple theme of construction. However the abilities of the non-passerine birds must not be disregarded. Although their nests are often relatively crude structures there are remarkable exceptions. They also include examples of dramatic contrast, from the neat and delicate structures, some only two centimetres in diameter, of felted plant down and silken webs fabricated by the hummingbirds, to the clumsy platforms of sticks collected by pigeons. The latter have even been known to make their nests from piles of discarded nails and the nesting ability of the pigeons is well-expressed in the old English nursery rhyme:

Coo, coo, coo
Me and my poor two
Two sticks across
And a little bit of moss
And it will do, do, do

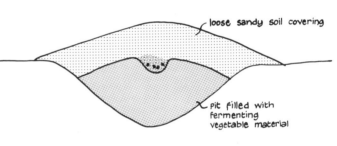

loose sandy soil covering

pit filled with fermenting vegetable material

Section of a mallee fowl nest. The eggs are laid, one a week, on decaying vegetable matter which gives off heat and incubates the eggs. The covering depth of the mound is altered by the male as he removes or replaces sand to maintain a constant temperature at the egg chamber.

Far left The hanging nest of the furous-fronted thornbird, *Phacellodomus rufifrons*, from Venezuela. The nest contains about four chambers with separate entrances and though only one pair of birds nests in it, usually in the lowest chamber, half a dozen or so birds may sleep there each night.

Left The cave swifts of Malaya and Indochina are able to produce strings of gelatin-like saliva which harden in the atmosphere. They use this material to construct their nests and thereby avoid having to find building materials. Unfortunately man harvests these nests to eat. They are knocked down from the cave walls with long poles as soon as the birds have finished building them.

Some of the ingenuity and great variety of
construction in birdnest-building is shown in
these examples:
1 The bracket-shaped nest of *Collocallia lowi
robinsoni*, a swiftlet of Thailand. This nest, about
5 cm long, is made of vegetation and moss mixed
with saliva and glued against a cliff face.
2 The nest of the little hermit, *Phaethornis
longuemareus saturatus*, of Costa Rica, which is
about 25 cm long, is built by the female from
cotton, moss, lichens and spiders' webs bound to a
slender palm frond.
3 The green broadbill, *Colyptomena viridis*, of
south-east Asia, builds its 38 cm long nest slung
from creepers on tall plants, often suspended
over water.

1

2

3

4

4 This large nest is constructed by the yellow-winged cacique of Mexico, *Cassiculus melanicterus*. The overall length is approximately 1 metre.

5 A typical hummingbird's nest, made by the white-necked jacobin, *Florisuga mellivora*, of Central America and about 22 cm in length.

5

6

7

6 The black-necked weaver, *Ploceus nigricollis*, is an East African species and weaves a retort-shaped nest measuring 22 cm in length, which hangs from the tips of the foliage.

7 Also suspended from drooping branches is the bulky nest structure of the cinnamon becard, *Pachyramphus cinnamomeus*, a bird of the Central American forest. The nest has a concealed side entrance and measures about 43 cm in length.

Despite their crude attempts at nest-building pigeons are nevertheless a prolific and successful family. They are found throughout the world in both temperate and tropical regions and have become an integral part of the urban scene. Another order of birds in the non-passerine group has also become closely acquainted with human societies. These are the Galliformes which include the pheasants, turkeys and common domestic fowl. They are all ground birds and though some may roost at night they all nest at ground level.

One small family among this order has developed the remarkable habit of incubating its eggs artificially. These are the shy and rather solemn megapodes, found in New Guinea, Indonesia and parts of Australia; they comprise the scrub fowl, the brush turkeys and the mallee fowl. Their nesting behaviour has earned them the general name of 'incubator birds'. The scrub fowl practise the simplest form of artificial incubation by burying their eggs in black volcanic sand and leaving them to hatch by the heat of the sun. On some of the Solomon Islands they are also able to make use of the heat from the volcanic steam which percolates through the soil. Other species, living among jungle vegetation, bury their eggs inside soil heaps raked into mounds up to 5 metres high and 12 metres in diameter. In locations where the shade is particularly dense vegetable materials are included in the mound and heat is generated by the rotting vegetation inside the nest.

The nest mounds built by the brush turkeys do not display such variety. In general they are restricted to the forested areas of New Guinea and the eastern coast of Australia. In these warm and moist conditions where vegetation rots easily and quickly they build mounds, usually a metre or so in height, composed almost entirely of plant material scraped from the forest floor. As it begins to ferment the male regularly inspects the mound probing inside to test the temperature with his tongue and controlling the heat level by removing or adding material. Not until conditions are correct does he allow the female to dig a tunnel into the mound and lay her eggs. For the next two months the male guards the mound, checking the temperature and adjusting it to suit. When the chicks hatch they dig their way out unassisted, and are independent from that moment.

In the semi-arid scrublands of Australia the mallee fowl are faced with difficult conditions and have been forced to devise complex procedures to ensure successful incubation. The air is so dry that the sparse vegetable material which is available will not ferment and the temperature fluctuates widely between day and night. The mallee fowl therefore begin their mound-building early in the year by excavating a hole, about a metre deep and up to five metres wide, and in the winter collect every scrap of dead leaf within the area to place in it. After these have been dampened by light showers of winter rain the mallee fowl hurriedly add a covering of sandy soil to protect the leaves from the dry air. The moistened vegetable material thus begins to ferment and generate heat. By early spring the rate of fermentation is increasing rapidly and the male has to work hard to control the conditions, digging down in the early mornings to uncover the eggs and allow cool air to circulate around them. As summer progresses and the heat from the sun begins to increase the male has the additional burden of increasing the layers of soil throughout the day to insulate the eggs from solar heat gain. But later in the summer, when internal fermentation has ceased and the heat of the sun is less intense, the temperature of the eggs may fall below the critical level of 30 degrees C. The mound is then uncovered early in the day to allow the heat of the sun to penetrate to the eggs ·and the bird gradually adds small amounts of warm soil until by late after-

The nest structure of the rufous-breasted castlebuilder, *Synallaxis erthrothorax*, from Central America appears to be an untidy collection of twigs. In fact it is well-planned and orderly. The bird's bill points to the small turret of fine twigs which is the entrance. This leads into a hallway carpeted with fragments of cast reptile skin and opening out into the nest chamber. The eggs are laid on a mat of green leaves and are protected from above by a thatch of coarse material which sheds the rain.

noon the mound is again complete. Finally the eggs hatch and the young birds emerge from the heap. The parents however have only about one month's rest before it is time to begin planning a new nest mound in anticipation of next year's brood if they are to defeat the harsh conditions of the Australian scrublands.

The swifts are another group of non-passerine birds which have evolved specialised nest-building techniques. They never voluntarily come to the ground and collect their building materials on the wing. This restricts them to wind-blown feathers and scraps of straw but they are able to supplement these meagre materials by using their copious and sticky saliva. They glue their nests inside hollow trees or chimneys and under the eaves of human habitations. Some of the South American species build their nests behind waterfalls and many swifts build against the walls of caves. The palm swift of Africa has reduced its nest to a small pad of feathers glued to the underside of the swaying palm leaves. The eggs also have to be glued into place with saliva and are incubated by the parent bird clinging upright to this platform. Another extreme example is the edible nest built by the cave swifts of the Far East. Their nest is composed entirely of saliva. They draw the thick mucoid secretions out of their mouths in strings which gel in the atmosphere and coagulate to form the small white nest cups.

The swifts' edible nest is certainly unique but perhaps the most ingenious nest built by this species is that of the scissor-tail or cayenne swifts of Central America. They build long tubular nests suspended from beneath a rock shelf or the eaves of a house. The nest tube, made of plant fibres and feathers matted together with saliva, is left open at the bottom and about halfway up the bird builds a small shelf where the eggs are kept. The cayenne swift's nest is remarkable in that it is equivalent to building a tunnel in the air. It is built by addition rather than subtraction of material which, as we have seen with such examples as the kingfishers and jacamars, is the normal method of making a tunnel nest.

Finally in the non-passerine orders we find one more example of hole-nesting which is worthy of comment. This is the species of tropical hornbills, and though they nest in ready-made tunnels such as hollow trees or woodpeckers' nests, they make interesting modifications to their nesting sites. Having chosen a suitable niche the female enters and seals herself inside. The entrance is plastered up with dung and mud brought by the male until only a narrow slit is left. The plaster dries to form a solid hard wall and the female has to be fed through the narrow aperture during the entire incubation period, which is an arduous time for both parents.

Nearly all birds take great care not to soil their nests and the hornbills are no exception. Scavenging insects inside the nest help to keep it clean and the female hornbill throws away any pieces of indigestible food. She is also able to achieve accurate high-velocity defecation through the narrow entrance, and it is claimed that the young birds pick up their droppings and deposit them outside. Removal of the faeces is normally associated with those birds which hatch inside a cavity, such as the hornbills, or in a nest made of delicate materials. This type of construction if clogged with soiled material would lose its insulating properties, and since the newly hatched birds are unable to regulate their body temperature this could mean a disastrously high mortality rate. Many of the passerine birds, which construct or line their nests with very fine materials, have developed a physiological adaptation to assist in nest sanitation. The nestlings of many passerine species discharge their droppings neatly packaged in a gelatinous capsule. The parent bird is thus able to remove the droppings easily and effectively. Obviously natural selection therefore favours those species which keep their nests clean.

A male striated weaverbird, *Ploceus manyar*,
building his retort-shaped nest in a swampy reed-
bed, near Delhi in northern India.

Despite their plant-like appearance the fan-worms
are animals. They have the ability to build
protective tubes of coagulated sand grains around
their bodies. The radiating plumes which the
worm spreads out from its tube are used both for
breathing and for catching food.

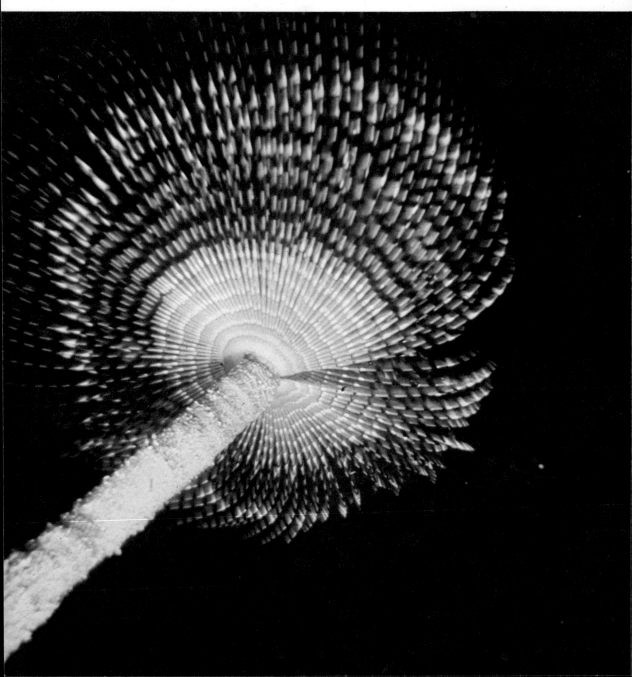

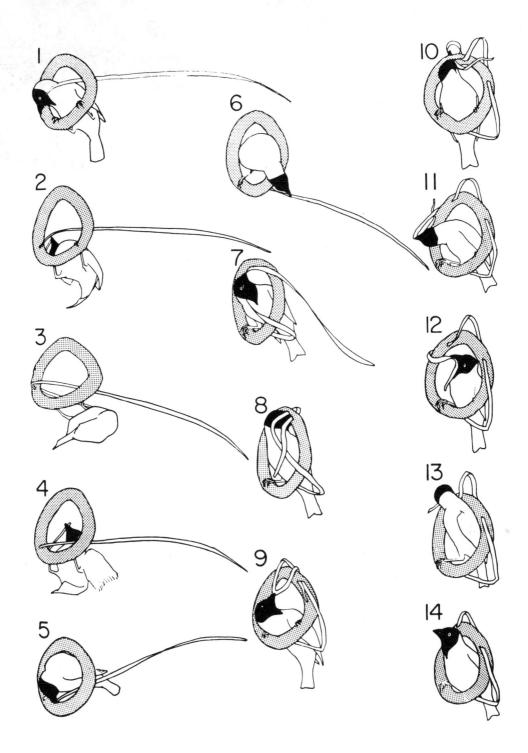

A typical sequence of movements by a male village-weaverbird as he weaves a single strip of elephant grass into the ring frame of his future nest.

Successive stages in the construction of the pensile nest built by the sulphur-rumped myiobius, a small bird of the central-American tropical forest. The nest, suspended from the end of a slender twig, is built entirely by the female and may take three weeks or more to complete. The originally continuous tuft of fibrous material is hollowed out from below and the fibres of the side wall are spread apart to form the egg chamber.

Right Successive stages in the building of a nest by the African village-weaverbird. While constructing the nest the bird always perches on the lower half of the initial ring frame. It also consistently faces into the nest, using its own body length as a module.

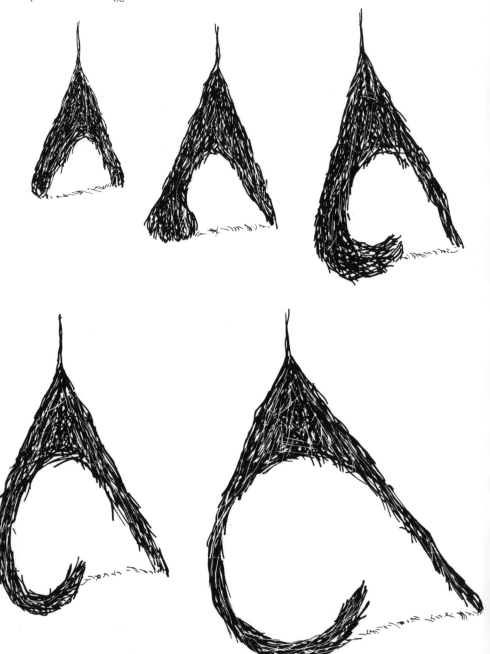

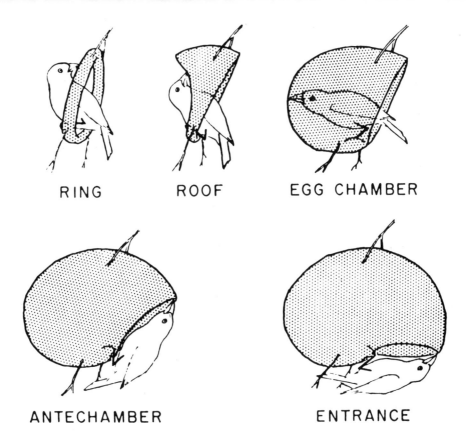

RING   ROOF   EGG CHAMBER

ANTECHAMBER   ENTRANCE

There are, as one should expect, exceptions to such a rule. Some of the tropical American cuckoos share a common nest which becomes extremely filthy as the chicks develop. But it is probable that these species are evolving changes in their nesting behaviour and any adaptations in nest sanitation are bound to lag behind such developments.

The passerine order of birds however have certainly found success in the system of natural selection and have evolved more rapidly than any other order. There are over 5,000 passerine species. They are all land birds, none of them particularly large, and include the most accomplished mimics and musicians. They tend to be active and intelligent; indeed the crow, a member of the passerine species, is credited by many ornithologists as being the most intelligent of all birds.

Although most passerines are small birds this does not restrict them to building small nests. A classic illustration of this is shown by one of the South American spinetails, a bird no larger than a sparrow, which builds a nest so large and heavy that though they may begin building in a branch some five metres high, the boughs bend down under the weight until sometimes the nest is resting on the ground. As one might expect there is a native legend concerning this peculiarity, which claims that the spinetail, by its song, persuades all the other birds in the forest each to bring a stick for its nest.

The spinetails are members of the ovenbird family, and this group has the most varied nest architecture of any bird family in the Americas, if not in the world, and illustrates the diversity of nesting behaviour among the passerines. The ovenbird family includes species which nest in open cups, in underground tunnels and in holes in trees. Some build pendant globes of green moss and other materials with an entrance at the bottom, but most build large and substantial domed nests. The cachalote spinetails build enclosing structures each large enough to hold a bird the size of a turkey and strong enough to bear the weight of a man. Another of the spinetails, the black-faced,

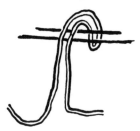

LOOP TUCK

SIMPLE LOOP

INTERLOCKING
LOOPS

SPIRAL COIL

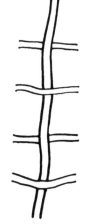

SIMPLE WEAVE

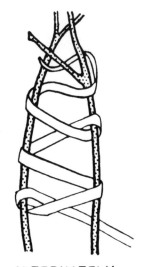

ALTERNATELY
REVERSED WINDING

HALF HITCH

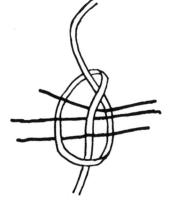

OVERHAND KNOT

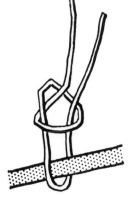

SLIP KNOT

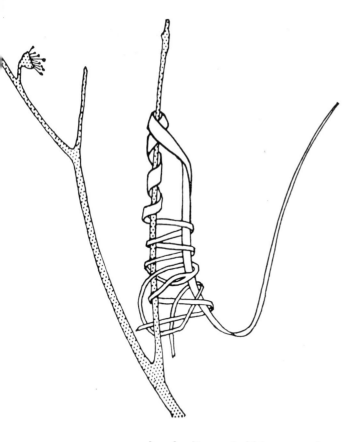

Far left and left Diagrams illustrating some of the elaborate weaving and knot-tying employed by weaverbirds in the construction of their nests, and a drawing showing the initial strip in the weaving of a new nest. The grass strip is first coiled about the twig, then alternately threaded between the dangling part of the strip and the twig. To become so accomplished young male weaverbirds spend much of their time weaving strips of grass around twigs or practising knot-tying, apparently simply for amusement.

uses mud and saliva to hold leaves and grasses together in a domed nest constructed so that each piece overlaps and forms a roof impervious to the rains.

The firewood gatherer, an Argentinian species of the ovenbird family, constructs a solidly built nest almost a metre deep using large sticks. These birds however seem to be ill-equipped for handling such unwieldy materials, and because the sticks they use tend to be larger than would seem necessary, and the birds are very weak flyers, building the nest is a laborious business. They frequently drop the stick in their struggles, or a projecting twig knocks it from their grasp, but it is never retrieved for the birds are not strong enough to fly vertically with such a load, and they do not seem to have enough intelligence to carry the stick away from the tree and then return at an oblique angle.

The herculean labours of the firewood gatherers result in a very secure nest which has an entrance at the top and a spiralling passageway leading down to the egg chamber lined with wool and soft grasses. After completing these labours and having built such a fine nest the birds will then sometimes demolish the structure and set out to build another. Even if they do not carry out this puzzling and seemingly suicidal action their nest is frequently taken over by some other species who drive them away.

The true ovenbirds and those which are responsible for the family name build stout domed nests of clay reinforced with horse-hairs and slender pieces of vegetation. Apart from giving coherence to the structure these strips of vegetation also assist evaporation of moisture from the mud and thereby tend to distribute the shrinking and cracking more evenly and safely. When completed the nest resembles an old-fashioned baker's oven. The rufous ovenbirds often construct their nests in the vicinity of human habitations, frequently on the roof of a house. The heavy structure is well-fortified and its planning makes it doubly safe. The entrance is deep and narrow, spiralling inwards to the egg chamber so that it is impossible to reach a hand into the interior. It almost seems as if ovenbirds might gain some kind of stimulation from building such durable structures so well. The firewood gatherers and spinetails do more building than would seem to be required, and the ovenbirds build a new mud nest each year despite the fact that their constructions remain sound for a number of years and are often taken over by swallows and other birds.

Enclosed nests of various forms are common among the passerines. Generally these are built of interlaced vegetable material and have a side entrance. Typical examples are the loosely woven nests of the exotic lyrebirds and the colourful pittas, but neater examples can be found among those wrens which build roofed nests. Some of the American wrens are unusual in that they also build another nest for roosting, and the habit of communal roosting in a special nest also occurs among some of the tits. These are all small birds, indeed their name is derived from the Anglo-Saxon word *tit* meaning a very small object. One member of this group is the cape penduline tit which gets its name from the type of hanging nest it makes. The nest of the penduline tit is perfectly built from fine and soft materials felted together to make a strong pouch. It is also virtually thief proof against marauding monkeys and other predators, not simply because of its strength but because of a cunning entrance arrangement. In effect the bird builds two nests inside one structure. One nest pouch is merely a sham: it has a large and obvious entrance but is left empty. The entrance to the real egg chamber incorporates a flap of material which is kept closed when the parent bird leaves the nest, thus disguising the access.

No less ingenious in the range of domed nest constructions are those built by the tailorbirds. Several species are familiar garden birds in parts of India and southern China and are extremely active and vocal. They have long and straight bills which they use as a needle for sewing one or two large leaves together as a cupped foundation for their nests. The birds pierce holes with their beaks near the edges of the leaves and draw them together using plant fibres, cocoon silk or scraps of cotton as a thread. Each stitch is made separately and the thread is knotted on the outside of the leaf to prevent it collapsing. Formed into the shape of a bag the leaf pouch is then stuffed with tufts of animal hair, cotton wool and thistledown. Three to six birds are hatched in the nest, which bears their increasing weight with no sign of the stitches slipping or tearing although the whole construction does tend to sag and swell at the seams.

Finally, in considering the nests of the passerines, the Afro-Asian weaverbirds deserve special mention, for they build some of the most interesting structures and display a greater variety of nest construction than any other family of birds. They build covered nests but some of the more primitive species construct a cup-shaped platform before adding a roof, which suggests that the original nest of the weaverbirds' ancestors was the more usual dish-shape built by the majority of birds. This form is suitable for holding birds' eggs and keeping them stable but the evolution of a domed nest offers other advantages. It gives additional protection against predators and heavy rains, and in the tropics against overexposure from the intense ultra-violet light. Domed nests built in trees have furthermore tended to evolve vertical entrance tubes which are effective protection devices against predators such as snakes.

The true weaverbirds all use fresh and flexible materials for weaving their nests, but other species which live in arid areas have had to develop thatching rather than weaving techniques. The available materials in such areas are too brittle for weaving and are only suitable for meshing together into a nest mass. Thatching techniques are used by the sociable weavers: every piece of grass is thrust into position, never pulled, but they build very large communal nest masses up to 6 metres long, 4 metres wide and 1 metre thick, the whole interlaced among the main supporting boughs of some large tree. A typical nest mass can easily contain more than 100 individual nests. Strong and thick branches are obviously necessary as supports for these nest masses, which when partially soaked by sudden storms must represent a considerable extra burden on the tree. Not surprisingly boughs do sometimes break off, bringing great chunks of the nest mass down with them. Nonetheless the birds have been known to continue feeding their young in a nest which has fallen to the ground.

The true weaverbirds make constructions by interlocking loops of flexible strips of materials. Generally the nests are fairly complex and dense. Some display extremely systematic and skilful examples of true weaving and a typical instance of this is found in the kidney-shaped nest of the common village weaverbird, a familiar resident in many settlements throughout Africa south of the Sahara. Construction is carried out by the male who gathers strips torn from elephant grass for weaving the shell of the nest. He bites across a part of the broad grass blade, then tears the strip away towards the tip of the long leaf and rips this free by flying vertically. The first strips for the nest are torn from the tougher central part of the grass blades and are wrapped around the fork of a twig to form a substantial foundation for the suspended nest. Additional long strips are interwoven so that they hang down in tassels on each side of the bird as he works. Grasping these two bundles of loose dangling strips in his feet the male then carries out a complicated feat of acrobatics, and working upside down weaves the tassels together one strip at a time until they are all joined into a ring.

This ring of grasses then acts as a platform and a module for the future construction work. The male builds a number of looped grasses into the ring to form the basis of the roof and side walls, and gradually as the bird works from his position on the ring, stretching as far as he can in all directions, the spherical shape of the nest naturally develops. To construct the hooded entrance in the side of the nest the male continues to work from the same position on the original ring of grasses but leans over backwards more and more until eventually the lip of the construction is in an almost horizontal plane.

While the male is building he strips almost all the leaves from the tree in the vicinity of his nest to give a clearer view of any approaching hawks and to help make the nest more conspicuous to potential mates. When all is completed the male advertises himself at the entrance and the females come to make an inspection. A female will demonstrate her acceptance by bringing bits of soft grass and feathers which she uses to line the interior. There is no further copulation after the eggs are laid and the male will leave his grass widow to start another nest and attract new mates. If the male is unable to attract a female after about a week he vents his frustrations on his nest, tearing it to shreds. In the same place he sets about building a new one, and may have to continue this procedure for a number of weeks, building perhaps as many as 20 nests. Each new structure fortunately is an improvement on the last and eventually the male builds a nest which is neat and sound enough to attract a mate.

The village weaverbird is typical of all the true weaverbirds in that its nest is an elaborately and beautifully woven structure. There is however a considerable range of techniques within the weaverbird group. Their abilities are also adapted to suit the materials available and the conditions of their environment. Thus those nesting in areas subject to heavy rainfalls build thick and compact ceilings into their nests while species living in dry areas may not construct any ceiling at all. The existence of a ceiling

The nest of the
dormouse is a rare
example of
mammalian
building abilities,
yet it rivals any
other construction
built by any of the
other animals.

is a characteristic feature of the spectacled weavers' nests. They weave a thick mat of material in the space above the nest and this is further protected by the overhanging palm leaves. The nest itself is also smooth and densely woven and because of the extreme slope of the roof any water which does fall on to the nest is quickly shed.

Probably the most proficient weaver of them all is cassin's weaver, which constructs a typically spherical nest with a long straight entrance tube hanging below. These nests are woven entirely from long narrow strips of palm leaves, the outer layer being compactly woven from strips about one millimetre wide with extra fine strips for the even more closely woven inner layers. Much longer strips of weaving material are used for the entrance tube and this part, unlike the main nest which is woven in a regular criss-cross pattern, is constructed with reflexive loops which spiral down to form a pattern of diamond meshes. The total structure is tough and resilient, displaying, in the classical tradition of good architecture, the attributes of commodity, firmness and delight: and all this with the most rudimentary of tools and the most basic materials.

These examples of birds' nests serve only as a glimpse of the great variety of forms and building skills to be found in the world of birds. Their abilities for constructing artificial shelter are matched only by a few other animals, notably man and some of the insects. The most accomplished builders in the insect world are the social bees and wasps, but their solitary relatives have also developed many interesting techniques. Although they do not build with such mathematical symmetry the solitary wasps and bees often construct cells which are perfectly formed. The potter and mason wasps in particular build beautiful small flasks of clay and sand grains glued together with saliva.

Outside the range of the birds and the social insects the examples of sophisticated nest-building are rare and scattered. And they are found in the most unlikely species. One of the common millipedes for example constructs an unusual egg chamber. Faecal pellets and earth are mixed together with saliva to form small pellets which the millipedes stack together to build a circular wall. The batch of eggs is laid inside and then another complete day is required to finish building the roof to complete the enclosing dome.

It is also remarkable that parental care is displayed by such animals as bullfrogs, earwigs and alligators. The nest of the alligator is very similar in principle to the mound built by the incubator birds, and the eggs are hatched by the heat generated from the decaying vegetable material which the alligator includes in her nest of sticks and mud. Racoons and many other animals are always ready to raid the nest and the alligator has to maintain a constant vigil to protect her eggs. The young alligators, like the new-born incubator birds, are left to fend for themselves after hatching though they sometimes spend their first winter with the female in her den.

Alligators are the only reptiles to build any kind of substantial nest, despite the fact that most reptiles lay eggs. But nest-building is much more common among the amphibians, especially some of the frogs. The blacksmith frog, like some other species of the tropical tree frogs, excavates individual pools out of clay in which to lay its eggs, while the nest of the grey tree frog of Africa is made from mucus exuded by the female which the male beats into froth with his back legs until it sets like a meringue. The nests are built in trees overhanging water and when the tadpoles have developed they wriggle their way out through the underside.

Apart from the primitive duck-billed platypus and the echidna no other mammals lay eggs. It is therefore only those mammals, the young of which are born in a relatively

What appears to be a jumbled heap of dead leaves
and broken branches on the forest floor is in
reality the result of a nest-building operation by a
mountain gorilla. Young trees and complete
branches have been broken down and pulled
around the body of the gorilla, enclosing a
comfortable mattress of luxuriant vegetation.

helpless state and which are not equipped with a marsupial pouch, which build some kind of shelter. The more accomplished nests are those constructed by small creatures such as squirrels and mice. The European red squirrel sometimes takes over the deserted nest of a crow or magpie as a foundation for its drey, but it is well able to build its own framework of twigs interwoven among the branches which it lines with moss, grass and hair. It is claimed that males and females build separate dreys for sleeping and that a larger nest is used for breeding.

The most perfect structure built by any of the mammals is the woven nest of the dormouse. It uses long summer grasses to build a spherical nest, often suspended between wheat stems, and lines it with plant fibres and chopped pieces of grass. Often the nest is camouflaged with leaves from the surrounding vegetation. A new nest is built for each successive litter and then in the autumn they move to their underground winter quarters where they store seeds and grain.

It is a temptation to make comparisons between the architecture of the human and the animal worlds, especially when discussing the building achievements of the mammals. At first sight it appears curious that so few of the other mammals, with their increased intelligence and dextrous abilities, are able to build any form of artificial shelter. In particular it might have seemed reasonable to suppose that the great apes, our nearest relatives, should be efficient and imaginative builders. Such comparisons however are made on difficult and dangerous ground, especially since modern man has domesticated himself and is now far removed in so many respects from the state of nature. But when we consider the way of life practised by stone age man and compare this with that of non-human primates we can more easily appreciate the lack of stimulus to develop building skills.

Prehistoric man, like the apes, was without agricultural skills. He had access to only a limited terrain for gathering food and had to move on when he had exhausted each locality. Mbuti pygmies in the African Congo still live by such a pattern, and primitive tribes living in Borneo spend no more than a few hours in building their temporary homes. The nomadic Bushmen of the Kalahari are also hunters and gatherers, taking advantage of natural shelter in caves, beneath overhanging rocks or building temporary screens of branches tied together with bark fibre and covered with grass held in place by chunks of wood. The floor is usually scraped out a little and strewn with grass to make a bed. Daytime sleeping places are also made, for the Bushmen, with only a thin blanket as covering, do not sleep well or for long during the night, and often catch up with their sleep during the day. Similarly the building activities of the great apes are limited to constructing sleeping nests and crude shelter for siestas.

Some impression of the attitudes which develop through living a wandering life in the jungle are revealed in the language developed by the Mbuti pygmies They refer to the agricultural villagers as animals and cannot comprehend why they should choose to live in such a noisy and diseased environment. In contrast the pygmies often use the word 'Father' or 'Mother' when referring to the forest which to them symbolises coolness, tranquillity and social harmony and provides nourishment, joy and security. It is impossible to discover what attitudes, if any, gorillas may have to their forest homes, but their way of life in a similar habitat mirrors that of the pygmies in many respects. The gorillas roam the Congo forests in small family groups, settling down as dusk overtakes them and building a bed for the night. They seem to prefer to build high but because of their weight they are often forced to build on the ground. In this case their nests are made simply from vegetation gathered together and placed to form a ring around their bodies. The nests regularly disintegrate in the night and indeed it is difficult to see how the gorillas benefit from building nests. They apparently do not provide any sort of shelter or insulation, but all the great apes have an instinct to gather some sort of material under and around their bodies before lying down to sleep.

The sleeping platforms which the gorillas build in the trees have to be more substantial than those on the ground. They sometimes pull branches down to form a horizontal platform or they may bend a bamboo tree down to the ground and stuff the bushy top of the tree amidst any neighbouring branches to act as an anchor. Similarly they will sometimes pull down a leafy bush, break and bend a few twigs and settle down to rest in the untidy but obviously comfortable heap. The branches are pushed and pulled into place, enmeshing together, and the ends are bent and twisted to form a ring around this platform. In an attempt to keep the whole construction together for a while the gorilla gives the twigs a sharp twist inwards so that they can be tucked into place.

The chimpanzees' nests are very similar to those of the gorillas and are also built each evening. But the chimpanzees are able to build much higher, anything from 5 to 50 metres above the ground, and their nests are quite well scattered among the troop. Nest construction is a simple process rarely occupying more than five minutes' building time. The foundation of the nest is usually a forked branch but if a natural site is not available or not suitable the animal splits a thick branch. The chimpanzee then begins to pull the surrounding branches down and towards him until they break. These are intertwined into the tree to form a rudimentary platform which is padded with leaves and twigs. Smaller branches are ripped off, one end forced into the nest structure and then bent to form a rough circular rim to the nest. Finally any protruding twigs and bunches of leaves within arm's length are added to the interior lining of the nest.

The amount of tree destruction by the great apes when building their nests is considerable. This behaviour may be linked with their apparent urge to destroy and break trees and bushes in their games just as birds often play with sticks and grasses. Again like the birds the apes seem to develop their building skills with practice. Orang-utans display nest-building behaviour at the age of only 18 months, crushing leaves and twigs into a crude platform for sitting on; these early attempts at building are carried out as play. As the animal gets older it spends more time in nest-building. By the time it is two years old the orang-utan may be able to build a complete nest structure in the sheltered fork of a tree usually between 7 and 25 metres above the ground, and when they are accomplished builders they always leave a branch sticking out as a lever for swinging themselves up to the nest platform.

Some years ago an orang-utan in London Zoo escaped by force from his barren cage. In a nearby tree he built a nest and settled down for the night. Next morning without any coaxing he returned to his cage and his breakfast. But the animal had made its point and now all the apes at the zoo are provided with wood wool for nesting. The material may be a poor substitute for branches and bushy leaves but it is a step in the right direction.

CHAPTER FIVE

Building Underwater

Life undoubtedly originated in the seas and although a variety of luxuriant fauna has made successful invasions of the earth, the land is in many ways a less favourable habitat than the water. Terrestrial animals have had to develop methods of supporting their own weight against gravity and they are faced with the perpetual problem of obtaining moisture. Special adaptations are necessary for reproduction outside the aquatic environment and for controlling the problems of evaporation. Furthermore the medium of air carries less food than water and is liable to greater and more rapid fluctuations of temperature.

Many of the building activities which land animals have to carry out are thus avoided by those animals which live underwater, and a few land animals have reversed the evolutionary process and returned to a semi-aquatic life. This is evidenced by various beetles who are able to carry entrapped bubbles of air with them under water for hunting trips. The water spider also traps bubbles of air in its thick covering of hairs to enable it to breathe underwater. This small supply however is insufficient for the spider's needs so it builds an oxygen reservoir for itself. The spider weaves a closely meshed horizontal web among plant stems, and ferries small bubbles of air from the surface to be released underneath this nest. As the bubble increases with repeated additions of air it begins to strain against the net and forces itself into a dome-shaped tent. Inside this tent the spider can sit and wait to dash out for passing prey. Mating between the water spiders takes place inside the female's tent, despite the fact that the male is almost twice her size, and the eggs are sealed off in the ceiling of her tent with a curtain of silk.

The prey of the water spider includes such small animals as water fleas and gnat larvae. Many flying insects lay their eggs in water and the soft-bodied larvae, though often carnivorous, have to shelter themselves against attack from larger animals. They

The Malayan soldier-crabs live between the tide marks on coastal sand flats. As the tide rises each crab buries itself in a bubble of air until the waters recede. Small pellets of sand are first pushed outwards and upwards until a dome has been formed to enclose the crab, which then travels deeper by scraping sand from the floor and plastering this on to the ceiling.

Below Another species which frequents ponds is the fishing spider, which builds a floating pad of dry leaves and debris tied with silk. The spider sits on its raft, ferried by the breeze, and hunts for small fish and aquatic insects as they come to the surface. Running swiftly over the water the spider secures its prey, sometimes even crawling down the stem of a plant in underwater pursuit, and carries it back to the raft.

Bottom Small bubbles of air, trapped amongst their body hairs, are ferried underwater by water spiders to form a reservoir between the stems of water plants. This oxygen tent is used as a lair for hunting small water creatures.

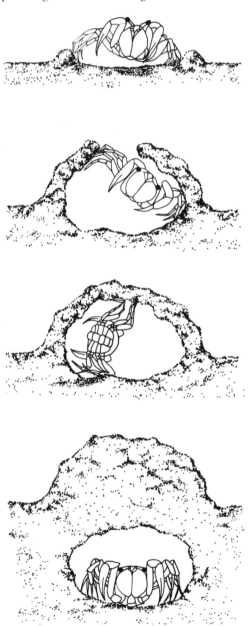

build a protective case around their bodies, in the form of an open-ended tube, from sand particles and bits of debris either cemented together with mucus or wrapped in silk. It is interesting that those living in turbulent waters build the most sturdily constructed cases, for animals which build underwater are faced with two main objectives: either protecting themselves against attack from other animals or obtaining shelter from violent water currents.

Waves pounding on rocks have tremendous force, as anyone who has sailed or surfed will readily appreciate. Animals living in such a habitat have to adapt to these conditions. Like the barnacles they are often protected by thick, heavy shells and, as with the limpets, they have tremendous gripping powers. Limpets also have a conical-shaped shell which offers a minimum surface area to the waves, and many other aquatic animals have evolved a body structure which helps them to live in swift-flowing waters. The flattened shapes of many swift-water animals present minimum resistance to the currents and further assists them to seek shelter under stones and in crevices. Clams bury themselves to avoid the current, settling down in an oblique position with their posteriors upstream.

A large number of aquatic animals seek shelter by burrowing, and it is not only the soil on the land which is richly threaded with burrowing worms. There are freshwater worms in the bottom of ponds and ditches. Many species live in the silt and mud of estuaries. They abound in the clean sand of the seashore, and there are also true marine forms. Like their terrestrial counterparts all these worms are soft-bodied creatures, generally cylindrical and perfectly adapted for burrowing.

The polychaete worms are almost entirely marine creatures and although many of them, such as the bright green or red ragworms, are free-ranging, crawling through mud and weeds in search of food, they also include many sedentary types which build permanent burrows and rarely if ever leave their shelter. The most familiar polychaete is the lugworm which is dug up from the seashore and used for bait to catch fishes. The lugworm lives in an L-shaped burrow with its head pointing to the blind end of the cul-de-sac. It is a hard-working animal but burrows only slowly. In short bursts of activity it agitates the face of the tunnel causing displacement of the sand above and forming a small funnel-like pit to cave in on the surface. Minute creatures and sedimentary deposits gather in this pit and slowly sink through the sand as the worm continues to vibrate the soil. Eventually they emerge inside the head of the tunnel and here they are eaten. Their remains, along with digested sand, are ejected from the tunnel in the form of a cast. This takes place once or twice an hour and it is the accumulation of these casts on the smooth shore which betray the presence of the burrow and its occupant.

The lugworm is an inoffensive though vulnerable animal, but some of the polychaete worms have developed the ability to build protective tubes in the sand. The beautiful peacock worms which gather like miniature forests in the lower regions of the seashore build tubes of sand grains around themselves and anchor them deeply in the substream. On a similar principle the sand-mason builds itself a ragged-looking tube though it uses such a variety of materials in its construction that it often appears to be in a state of partial collapse. The long and flexible tube built by the owenia worm is constructed in an unusual manner, using grains of sand and fragments of debris fixed in such a way that they overlap like the scales of a fish or the tiles of a roof. Its tube is about 50 millimetres in length, but that of another species, megalomma, builds a tube of sand and gravel some 300 millimetres long, and weights the base with a conglomerate anchor of pebbles and shell fragments. Like the tube of the peacock worm this projects about 20 millimetres above the surface and is closed off by the worm when it retreats from the receding tide.

The function of the tubes built by the polychaete worms, like those built by the caddis fly larvae, is to protect their soft bodies. All that is normally visible is the feathery crown of tentacles, often brightly coloured, which fan out and filter food particles into the mouth. The pectinaria however is a polychaete worm which, although building a substantial and regularly formed tube from sand grains and shell particles, lives upside-down and digs into the sand for food.

On this simple principle of building a tube for living in the polychaetes have evolved a wide-ranging variety of methods. Some make gelatinous tubes of coagulated mucus which are so fragile that they collapse if disturbed whereas other species build groups of tubes so strong and thick that they form large honeycombed reefs of reconstructed sandstone reinforced with shell fragments. Prominent structures are also made by those polychaete worms, normally living on the seashore, which deposit calcium carbonate in dense and twisted masses to resemble heaps of petrified spaghetti.

The animals which are normally associated with providing a calcareous armour for themselves are the molluscs. It is strange that some of them, although protected by strong shells, burrow themselves into rocks. And it is their protective shells which they utilise for this activity. The bivalve piddocks for example oscillate on the ball joint of their shells and spend their entire lives slowly grinding away the rocks, gripping with their muscular foot and changing position as they progress to produce a smooth rounded tunnel. Having started their relentless burrowing they often continue in a steady straight line, passing through their neighbour's tunnels and occasionally through a neighbour's body.

Most of the tunnelling bivalves seem to prefer rocks, but the notorious ship worm eats wood and causes great damage by its tunnelling activities. In San Francisco during 1919 and 1920 ship worms brought about the collapse of wharves and docks causing damage valued at $20 million. Despite its common name it is a mollusc not a worm, and it tunnels in much the same manner as the piddocks except that its secretions give a calcareous lining to its tunnels. It scrapes into the wood with the serrated edges of its shell but slowly and continually turns through a circle so that the diameter of the tunnel is wider than the shell itself. The rotary shield which is still used today for tunnel boring was designed by Sir Marc Isambard Brunel after studying the tunnelling method of the ship worms.

Burrowing both for food and shelter is practised by many other aquatic animals which, like some of the molluscs, seem at first sight unlikely candidates for such an occupation. Brittle stars for example excavate a burrow in which to conceal themselves and lie with just the tip of their long arms projecting a little way above the surface to catch their food supply of minute organisms. Some of the sea anemones are also powerful burrowers. The peachia anemone, which is named after the eighteenth century biologist Peach, and also happens to be peach coloured, occupies a burrow up to 300 millimetres deep and retreats to the bottom of its pit if disturbed. Halcampa is an anemone which often burrows beneath rocks and stones, while cerianthus lines and strengthens its burrow with a tube of coagulated sand grains.

The simple body structure of the invertebrate anemones is shared by a great number of marine animals. The body is merely a cylindrical wrapping around the gut, often with just a mouth and some means of catching food at one end and an anus at the other.

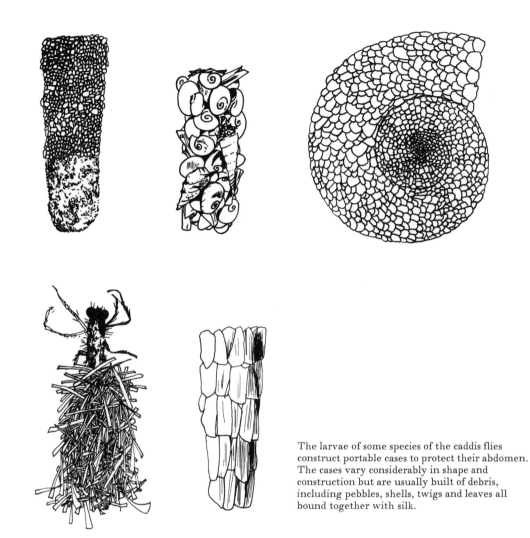

The larvae of some species of the caddis flies construct portable cases to protect their abdomen. The cases vary considerably in shape and construction but are usually built of debris, including pebbles, shells, twigs and leaves all bound together with silk.

Many of the animals with such a basic kit of parts settle into a burrow among the sand or mud. In this way they protect their bodies and leave only the mouth parts exposed. Some of these animals, such as the sea cucumber, improve this system to some extent by making a U-shaped burrow so that they can defecate from the opposite end and thus avoid fouling the burrows with their own waste.

Many creatures of the seashore are also adapted to a burrowing existence. The Malayan soldier crabs live on the shore between the tide marks and bury themselves inside a sealed chamber of air when the tide rises. The crab first scrapes a shallow depression, pushing small lumps of wet sand outwards and upwards to form a circular wall. From this foundation it builds a domed structure over itself, scraping pellets of mud from the floor and plastering them together at the extremity of its reach. When it is enclosed in its bubble the soldier crab begins to bury itself, using the same techniques,

1

2

3

4

The sticklebacks are rare examples of fishes which build nests. The 3-spined stickleback, living in open conditions in fast-flowing water, digs a shallow pit in the sand as a place to construct its nest. This is a recent evolutionary development, and all other stickleback nests, typically constructed of scraps of weed glued together, are placed amongst the branches of weeds.
Shown here are the nests of (1) the 3-spined, (2) 4-spined, (3) 5-spined, (4) 10-spined and (5) 15-spined sticklebacks.

5

by taking sand from the floor and plastering it to the ceiling, until it is safely submerged with its reservoir of oxygen. Here it waits in safety until the tide recedes.

More familiar burrowers on the seashore are the cockles and mussels. Their flattened shape is well adapted to burrowing and the ridged patterns of the cockle's shell help it to remain steadily anchored in the shifting saturated sand. At about a finger's depth below the surface of the sand the cockles gather into solid layers at an estimated population of 10,000 to a square metre. For some unknown reason a whole bed of cockles sometimes will suddenly up and move off to fresh pastures. They literally plough their way through the sand but their powerful foot is also occasionally used for shooting off in small and generally not very effective jumps.

The razor shells are also efficient burrowers. They pull themselves down into the sand and between each downward pull the shell is forced open to hold its position by lateral pressure, while the muscular foot extends to make a grip for another pull. With swift sharp movements they disappear very rapidly.

Of all the 80,000 different species of molluscs the most developed are the active and intelligent squids and octopuses. It is unusual to find responses of sympathy or endearment towards octopuses. Jules Verne easily transformed them into monsters, and skin divers seem to hunt them with even more relish than they do fish or other marine animals. It is perhaps debatable whether the form of an octopus might appeal to the finer senses but when it is engaged in the serious business of love-making an octopus is a difficult thing to despise. They emulate our own behaviour in such obvious manner, wallowing in great, tentacle-enmeshed embraces, dancing with each other and indulging in exuberant petting sessions. However at most other times the octopus tends to be a shy creature although when attacked or taunted it can display the most appalling rage. But generally they prefer to remain hidden among the rocks and corals.

The octopus seeks out a suitable recess for shelter and then sometimes closes the entrance with stones, large pieces of rocks and hundreds of empty shells. The stones and shells are gathered locally and carried back to the building site to be stacked on top of each other until there is just enough room for the octopus to enter. This entrance can be incredibly small for the animal can squeeze its body through a hole many times smaller than the diameter of its body, as many aquarium keepers have discovered to their loss.

Kept in aquarium tanks with no recesses or available building materials the octopus will snuggle tightly into a corner or try to bury itself in the gravel. Because they are sometimes kept in the wrong type of environment they also develop aberrant behaviour, eating their eggs and even killing themselves from self-inflicted wounds. The desire to enclose themselves with solid surroundings is overwhelming and sometimes leads to their doom. Fishermen let down pots on ropes and these are soon taken over by octopuses. They refuse to vacate such a perfect apartment even while they are being hauled up to the boats. Investigations of a Greek ship which had sunk off Marseilles some 2,000 years ago revealed that the cargo of fat-bellied wine jars had all been taken over as an estate for octopuses, and they have also been known to take residence in human skulls. They can however be ingenious in making their shelter. One den discovered by Jacques Cousteau in 1953 consisted of a lean-to structure with a roof made from a large flat stone propped up by stones and a brick at one corner.

Some species of female octopuses lay strings of about 150,000 eggs and hang them from the roof of their lair, carefully and intricately intertwined. During the brooding period the female guards the eggs and keeps them clean and well-aerated by funnelling jets of water on to them. The eggs take several weeks to develop and very often the mother dies of starvation during this period. Those species which do not guard the eggs hide them in a safe place: they have been found in bottles and casks and even in the fuel tank of a submerged aeroplane. On the sandbanks of Florida small octopuses regularly lay their eggs inside oyster shells, staying inside and clinging to both halves to keep the shell closed.

Octopuses spend a great deal of their time exploring their immediate surroundings by touch; anything which projects is pulled and if it moves it is gathered up in front of the lair. In aquariums any apparatus which can be broken apart is dealt with and the average life of a thermometer for example is about 20 minutes. It is doubtful whether the octopus builds a wall in front of its den to hide itself from view, for it has been shown that they do not differentiate between transparent or opaque objects as building materials and they seem to nestle as readily between sheets of glass as between slates.

None of the fishes build any such shelters for themselves. Generally they try to avoid their enemies by fleeing, or they are protected by a bony armour, as is found in trunk fishes, or by some other device such as electrical charges or poisonous spines. Others avoid detection by camouflage or by hiding in weeds. But some fish build nests for their eggs. At its simplest this in only a hole scraped in the gravel but some species make use of bizarre nesting places. Bitterlings for instance deposit their eggs inside a living mussel, and the lumpsuckers lay eggs inside a crab's shell. Extreme examples are some of the tropical fish which carry their eggs and developing young in their mouths. After hatching, the young swim around the parent fish while all is safe but when danger appears they dash back inside the parent's mouth. Equally strange is the kurtus fish which keeps its eggs in a special cavity on its forehead.

A few fishes blow frothy masses of bubbles which are used as a floating nest for the eggs and thus, like the water spider, take special advantage of the physical properties of

Diagrammatic sections showing a razor shell pulling itself into the sand. Between each downward pull the shell is forced open to hold its position by lateral pressure while the muscular foot extends to make a grip for another pull. The animal burrows rapidly with swift and sharp movements.

Bottom Male 3-spined stickleback spawning in its nest. The male stays to protect the eggs and spends a great deal of time fanning aerated water through the nest as the eggs develop.

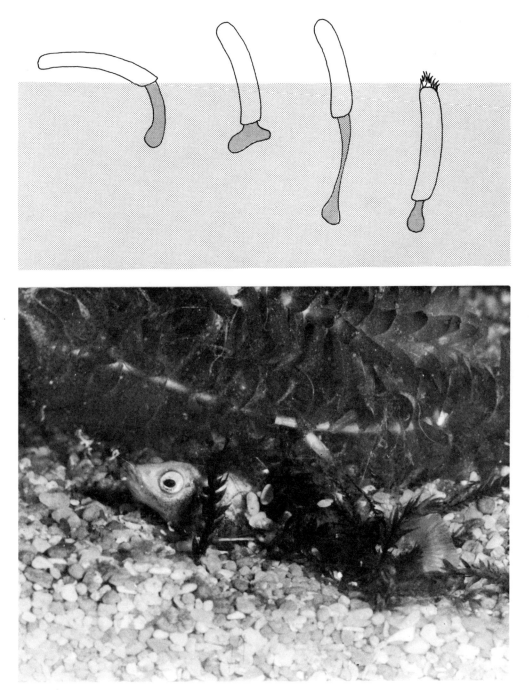

their aquatic environment. But perhaps the finest example of nest-building by any of the fishes is that of the three-spined stickleback. Certainly its nesting behaviour has received an extraordinary amount of attention from scientists, and studies of the stickleback's breeding cycle illuminate the importance of external stimuli as triggers for specific patterns of behaviour.

During the breeding season the male stickleback is a territorial animal and when he has established his territory builds a nest to attract a female. First he makes a shallow scrape in the sand and fills this with a collection of bits of weed which he glues together with a sticky fluid secreted by the kidneys. He then burrows into the nest forcing a tunnel through the pile of weeds. When the nest has been completed the male is ready to begin courting any passing female. If one responds she is led to the nest and the male indicates the entrance to the tunnel with his head, enticing the female to enter and lay her eggs inside the nest. Usually two or three females are induced to lay in the nest and each time the male slips into the nest tunnel after a female has left, to fertilise the eggs. The male stays to guard the nest and its eggs, fanning water through the tunnel to keep it aerated. When the young hatch a week or two later he continues to protect them, chasing after any that stray too far and carrying them back to the nest with his mouth.

The only example of a fish making any sort of shelter for itself rather than for its eggs or young is the African lungfish. During the rainy seasons the lungfish live in the flooded marshlands but later when the land dries out they dig down into the mud and roll themselves into coils. In this underground chamber the lungfish exudes a slimy coating around itself and this hardens to form a cocoon, sealing the fish from the dry heat of the land. Undisturbed the lungfish can live inside this casing for six months, waiting for the rains to flood the land again and soften its release. But during its fast it is vulnerable to attack from man. The airholes in the clay betray the presence of the fish and they are dug up and carried off, sealed in their own canisters, to market and the dinner table.

Animals burrowing in the mud of the seashore seek food and shelter at characteristically different levels. *Left to right* Lugworm, furrow shell, tellin, sandhopper and cockle.

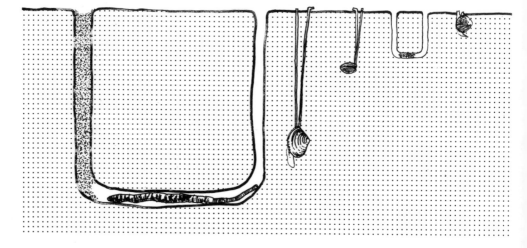

CHAPTER SIX

The Co-operative Builders

The destructive effects of man's society have had a devastating impact on many other community dwellers. The grass plains of the world tolerate large population densities: such as individual herds of bison in North America which sometimes composed over four million animals. But the ease and rapidity with which man decimated these populations with guns is a well-known and tragic story. The passenger pigeons were equally vulnerable to man because the densities in which they accumulated made them so easy to kill. Prairie dogs which in the nineteenth century almost certainly existed in billions and carved out towns in the prairies often covering several square kilometres have been severely reduced in numbers, though they are still numerous in parts of the south-western United States.

The advantage of the prairie dogs, which perhaps saved them from the fate of the passenger pigeon and the almost total extermination of the bison, is that they live underground and hide in their burrows at any sign of danger. Their townships form a social organisation which resembles man's society in many ways, and in comparison with, say, a European medieval township may even be superior in some respects. Infant mortality among the prairie dogs is extremely rare and their defence systems are virtually impregnable to all but modern methods of warfare. At all times there is at least one prairie dog on guard keeping a constant watch for danger, and at the first cry of alarm every prairie dog in the neighbourhood scuttles into the safety of its underground burrows.

Each social group occupies a defended town up to an acre in area and includes some 40–50 burrows. Some burrows are large enough to house up to 15 individuals in a family unit or coterie made up of one or two adult males, a few females and the young ones. The family defends its own burrow against intrusion by its neighbours and there is also a demarcation of the spaces inside the burrow itself. Branching off the main exit shaft are a number of horizontal tunnels, one for each inhabitant. The members of a new litter stay in the nest which is carved out at the end of the main tunnel and usually lies about six metres from the entrance hole. Excavated earth is piled around the entrance hole in a conical mound. This acts as a barrier against flash-storm floods as well as serving as a raised look-out post.

The viscacha, a rabbit-like inhabitant of the Argentine pampas, has adopted both a habitat and a way of life very similar to that of the prairie dog. Though they have long been the subject of an intense war of extermination they too manage to survive because they seek shelter underground. They also live in communities, of 20–30 individuals, and benefit from having sentries to raise the alarm if danger threatens. For this purpose the entrances to their burrows are closely grouped, and as with the prairie dogs the mounds which become built up around the burrows serve as both vantage points and protection from flooding. They clear the ground all around the entrances, cropping all the grasses and removing them, along with all the other odd bits and pieces they might find, to the entrance mounds to build them even higher.

Generally peace-loving and sociable the viscachas are nevertheless resentful of any intrusion into their burrows from their neighbours. It is the one action which they repel with intense fury and so strong is their territorial impulse that even when hunted by dogs a viscacha will not enter the burrow of another, except when faced with the most dire demands of survival. Despite their insistence on privacy they do however possess a true neighbourly spirit. If one of the burrows collapses for example or, as is often more likely, it is deliberately covered by a man and the buried animals are unable

The *Malacosoma* caterpillars often live socially as a means of protection against predators and, as shown here with *M. neustria*, the larvae of the lackey moth, they spin a communal web.

Left Relatives of the arboreal squirrels, the prairie dogs adapted successfully to life on the plains by seeking shelter underground. Their networks of underground burrows and nesting chambers once formed colonial townships over hundreds of miles.

Right A small group of rabbits were released in Australia in 1859. They rapidly multiplied having no natural predators to keep them in check in this new environment, and by 1950 their population was estimated at 1,000 million. The effects that these social animals have made is dramatically illustrated by this scene which is bisected by a rabbit-proof fence.

to extricate themselves their neighbours come and dig them out.

The wild rabbit is in many ways the European counterpart of the South American viscacha. A timid and apprehensive animal it lives only in large communities and although the burrows dug by one family of rabbits may be few in number the rabbits often combine with other families to form warren networks over hundreds of square metres. They have been introduced by man into many countries, where they have spread rapidly. Their extensive tunnelling and omnivorous diet have caused serious damage, especially in Australia where three pairs were let loose in 1859 near Victoria and within three years had multiplied into colonies numbering millions. The pregnant female rabbit makes her fur-lined nest in a shallow hole some distance from the family warren; this she camouflages with a covering of vegetation whenever she has to leave it. A family's burrow however is dug deeper into the ground and contains a central main chamber from which a number of tunnels spread out, leading to the open.

The rabbit has resisted all-out attack from man. In recent years the spread of myxomatosis has almost wiped them out in Europe but there are indications that once again they are on the increase. However it is only in the last century that rabbits have been such a pest, particularly in alien environments. In the Middle Ages they formed an important part of man's rations but now both their fur and their flesh are considered less valuable.

There has been a reverse situation with the beaver. Once distributed in hundreds of millions over North America from the tip of northern Mexico to the aspen limit of northern Canada, and over the whole of Eurasia from Lapland to Italy and from the

eastern states of Russia to England, the beaver is now limited to small and isolated parts of the world. The French pursued the beaver indefatigably throughout Canada. They began trading in 1603 under a royal charter and in 1669 the Hudson Bay Company joined in the lucrative slaughter with a charter from Charles II. The trading in beaver skins opened up the whole continent and was carried out on a vast scale. Today the beaver is protected by law in Europe and North America. It is becoming more numerous in eastern parts of the USA and has started building once more in Norway and Poland.

It is interesting that the beaver, when reduced to very small numbers in western Europe, seemed to lose its abilities for building dams. Many animals, particularly social types, often seem to lose the will to survive when their population is reduced below a critical level. This phenomenon is evident in a variety of species, including primitive tribes of men. It is indicative of the important and integral part which building plays in a beaver's life that this activity was the first to suffer under these pressures.

However, the beaver is not the constantly active and hard-working animal that he is so often made out to be. 'Busy as a beaver' is as much a misnomer as 'free as a bird'. Like all other animals the beaver works only when it is necessary, but on those occasions he does carry out tremendous tasks of labour, equal in relation to the building activities of the village weaverbird or the solitary mining bee. It is the large scale effects which the works of the beaver have upon the landscape which multiplies the significance of their labours. For the beaver, in setting out to create the environment to suit his own requirements, alters the face of the land. The creation of the English fens may have

1

These photographs illustrate the effect on the landscape created by some of the social birds.
1 Portion of an oropendola colony in Central America.
2 A rookery in England
3 Weaverbirds' nests on an African palm tree

been due to the destruction of natural drainage by the beaver which did not become extinct in Great Britain until about the thirteenth century. Similarly the destruction of the Pennine woodlands and the formation of the peat mosses of Lancashire have been attributed to the beaver.

In the USA many of the forest meadows which the pioneer settlers were able to cultivate had been created by beaver floodings. Although the beaver has disappeared from many great tracts of land its name is perpetuated in placenames scattered all over North America, such as Beaver Creek and Beaver Valley. In England it is remembered in Beverley and Beversbrook, and the early history of France records many allusions to the works of the beaver. There are also many traditions and legends of the American Indians concerning the beaver and it is significant that these are invariably connected with the creation of the world.

Beaver colonies are made up of a collection of family units, each comprising the parent pair, their annual litter of kits and the yearlings from the previous season. When they are two years old the young beavers either leave voluntarily or are driven out of the colony and wander off to set up another group. Unless they live in a natural pond or lake the first activity in founding a new colony is the building of a dam across a stream or river to create an artificial pool. They often make use of natural elevations in siting their dam but equally regularly choose an awkward site which has to be abandoned. The dams are built whenever there is an accessible supply of food and since the staple diet of the beavers is the saplings of willow, aspen and maple this also ensures that building materials are to hand. Normally the major part of the dam consists of branches of trees but occasionally, in swampy ground, the dam may be built solely of mud. The animals obviously take advantage of the nearest available material, and one dam found in 1899 in North Dakota was built entirely of coal taken from a nearby bluff. The usual construction of a dam is of felled trees and branches, sometimes of considerable size, dragged and floated into position and pushed into the river bed with the butt ends facing upstream. Mud, gravel and stones gathered from just above the site of the dam are placed on these timbers to weight them into position. Other layers of brush are often added and weighted down with mud. Sediment and debris brought down by the stream helps to block any cavities and more mud is added to the crest and to the downstream side to add extra weight against the pressures of the rising water. As the pond begins to fill up with silt the dam has to be raised and as the waters spread out it also has to be extended in length. The side extensions naturally follow the lie of the land causing the dam to become curved or angled.

The dams are built for a variety of purposes. Primarily they create a depth of water in which the animals can retreat from their enemies and in which they can transport and store their food supplies. But not all ponds are inhabited. Some are created in series along the river so that food supplies can be floated down to the main pond. Upstream dams are often built to control floodwaters so that they escape only gradually. The extension of the dams averages about 15 metres but it can vary from less than a metre to more than 1,000 metres, which was the size of one found on the Jefferson River in Montana. Such a dam would be the result of many generations of work. The dam is under constant repair and as it gradually silts up new dams become built on top of existing ones. Excavations have shown some to be 1,000 years old.

Beavers are also great canal builders, and although canals may be simpler to construct than dams they are much more difficult to site. They are excavated for a number of reasons, usually for the transport of logs to the dam site and as waterways in which

Left The hunting of the beaver opened up many parts of North America to the European capitalists. This detail from an early map of French Canada shows the fabulous workings of the beaver. An accompanying key explains the various activities in a mixture of fact and fiction:

'A   Beavers cutting big trees with their teeth
B   The carpenters who trim the long branches
C   Wood carriers
D   A group making the mortar for the dam
E   The clerk of the works, or architect
F   Nurses to care for sick workers
G   Labourers carrying mortar on their tails
H   A beaver worn out from working too hard
I   The masons who fashion the dam
L   Beavers consolidating the construction by beating with their tails
M   The domed house of the beaver, with one exit onto land and one into the water.'

The illustration also includes (*bottom centre*) a small section through a beaver's house, with a 'ground floor for the family, and an upper floor used as a wood store, their source of food'.

The beavers' lodge is built of sticks cut from the nearby woods and ferried to the site. The lodge contains a living chamber, just above water level, which has a ventilation shaft over it where the sticks have not been packed out with mud. In severe winters more mud is often plastered on to the outside of the lodge. The entrances are cut below freezing level. The dam is built up with stones and sticks and consolidated with mud and weeds. The construction of dams creates a lake for the beavers wherein they can safely build their lodges, travel unseen underwater, and store their winter food supply of cut branches.

Below left A beaver at work on its dam, photographed in Oregon. Having exterminated the beaver over great parts of the continent, the United States and Canadian governments are now actively engaged in conserving the animal in its remaining habitats.

Below The dams built by beavers often cause flooding on an extensive scale. After a number of years the workings may be abandoned and the waters slowly recede to leave a fertile valley meadow which at some time may once again be taken over by a new generation of beavers.

the beaver can travel with more security than on land. Varying in length from just a few metres to many hundreds the canals may also be built at two or more levels, each separated by small dams, the different levels of the system being supplied from streams at higher elevations. Small canals are sometimes built simply to divert the natural flow of a stream which then follows a new slope, but occasionally the beavers discover that they cannot divert the water in the required direction and have to abandon their diggings.

For winter protection the beavers build themselves a lodge. This is normally a half-submerged beehive-shaped structure. There are rarely more than four to each pond though sometimes they are built in the banks away from the water. The lodge is made of interlaced branches daubed with mud and always has some sort of solid foundation, either a submerged tree or some small natural island, but if no natural foundation is available the animals build one out of brushwood and heaps of clay. The sticks and mud are piled on top of this foundation to make a dome, and an air shaft is left between the sticks in the centre where no mud is placed. The large and well-insulated chamber inside the lodge is reached from underwater entrances below freezing level, which the beavers excavate by tunnelling into the lodge, gnawing off the branches as they progress. The lodge is increased in size by the addition of more material to the outside and by enlarging the chamber; small lodges sometimes merge to form one large structure. Large heavy logs are placed around and on top of the lodge to hold the construction together and in cold regions extra layers of mud are plastered to the outside. Snug and safe inside this impregnable lodge the animals do not have to expose themselves to the prowling wolves and other predators on the land, for the vegetarian beaver keeps stocks of wood inside the lodge and piled in heaps under the water.

Geraldus Cambrensis, travelling in Wales in 1188, described the works of the beaver which he saw, and though, as he said, the beavers' habitations are 'rude and natural without, but artfully constructed within', they did not match the fabulous stories brought home by many of the European explorers of the sixteenth and seventeenth centuries. Marvellous tales were expected from these early travellers and they did not disappoint their listeners. Thus they claimed that the beavers were incessant builders who used their tails as trowels to plaster their three-storied houses; that they had laws and a government; chose commanders who distributed tasks; posted sentinels to warn of enemy approaches; and exiled the idle. Much of these accounts was imagined, and a great deal more was misinterpretation of the beavers' way of life compared with that of humans.

Despite the abundance of such fantastic tales men were more interested in dead beavers than live ones. For 200 years beaver skins were the most important export of the North American continent and were the chief commodity of trade on the frontier. Lewis and Clark recorded seeing beaver ponds in Montana stretching away as far as the eye could see, and it has been estimated that at the beginning of the seventeenth century the beaver population of North America was in the region of a hundred million. By the late nineteenth century when beaver fur at last began to become unfashionable they were extinct over the greater portion of the land they had occupied.

CHAPTER SEVEN

Colonials and Imperialists

The tiniest animals on this planet exist in the nutritive environment of the aquatic world. They are all primitive forms of life, often found in the indistinct shadow between plant and animal form. Protozoa are living representatives of the first simple organisms which evolved as animals. The first to be seen were discovered by Van Leeuwenhoek, using a simple microscope, in 1674, and it is estimated today that there are some 250,000 species of protozoa. They present a bewildering variety of forms and behaviour but many of them display a mode of life which is echoed throughout the whole of the animal kingdom and which has proved to be an immensely successful adaptation. These are the colonial protozoa and a typical example of these is *volvox* in which several hundred individuals live together as one minute spherical cell. They breed in a remarkable manner. Individuals from the cell wall branch off and cluster in the jelly-filled interior of the ball where they multiply to form a spherical colony within the parental walls. Later they are released into the outside world but sometimes young colonies inside the parent colony will also develop their own offspring, resembling the constructions of a Chinese ivory ball. Despite their primitive state they represent a perfect example of true community life in which each individual is totally dependent on the others. This is a successful way of life for many species of animals; even man finds it difficult to live alone.

The formation of the *volvox* colony possibly resembles the manner in which individual cells originally gathered together to form multi-celled animals, the most rudimentary example of such a multi-celled structure being found in the sponges. In simple terms the body of a sponge is made up of an aggregation of individual cells which all join together in a fixed pattern. The surface of this body is made up of a network of star-shaped cells and between these are numerous small pores. Water is streamed through these perforations by the actions of certain cells which are equipped with whip-like flagella. The concerted thrashing of millions of these flagella drives the water into a system of tubes running through the body and these channel the flow to the exhalant vents. As the water streams into the pores any particles of food become trapped and these are then taken by the amoebocytes, which are special cells wandering about in the labyrinth of spaces between the star-shaped cells.

The amoebocytes have a variety of other functions and one of them is the building of the colonial skeleton which gives the sponge some rigidity. With few exceptions this is constructed either of a network of spongin fibres, as in the bath sponge, or from spicules of carbonate of lime. In the latter instance the spicules do not grow as an ordinary skeleton but are formed individually in the interior of the sponge and then carried into their required position by a group of amoebocytes. Not surprisingly this remarkable activity has attracted much attention from biologists, while the crystalline form of the spicules, which are tri-radiate structures inclined at nearly equal angles, have been closely investigated by physicists.

Recent studies have suggested that some sponges possess a primitive type of nervous system, but whereas some zoologists contend that sponges are colonies of single-celled animals others claim that they are not animals at all in the strictest sense. They are certainly a mysterious offshoot of evolution and although Aristotle was one of the first to recognise their animal nature this was not widely accepted until the nineteenth century.

Generations of naturalists were also preoccupied with the problem of classifying the corals. Réaumur regarded the calcareous structure of corals as being formed in the same way that bees construct their combs, so the expression 'coral insect' was introduced. John Ellis was nearer to the truth when in 1755 he wrote: 'I own I am led to suspect that by much the greatest part of these substances . . . are not only the residence of Animals, but their Fabric likewise; and serve for the purposes of Subsistance, Defence and Propagation, as much as the cells and combs fabricated by Bees and other Insects.' Linnaeus however believed the coral to be a vegetable which became metamorphosed into an animal when it flowered, whereas Rumphius, the seventeenth century Dutch naturalist, unable to unravel the mystery of the coral, had simply written off the oceans as a survival of primordial chaos.

In fact the true corals are colonial, carnivorous animals, closely related to sea anemones and jellyfishes. In clean and shallow tropical waters they coalesce in millions to form coral reefs in which individual polyps live together, each anchored by small cup-shaped deposits of calcareous secretions. By building reefs the corals have played as important a part in the history of the seas as the earthworms have on land. Three main types of reef are built: the fringe reefs, the barrier reefs and the atolls. Fringe reefs spread out from the shore, growing only along the projecting edge of the reef, leaving behind a deposit of dead corals. The Great Barrier Reef, though in serious danger from both the predatory crown-of-thorns starfish and from prospectors who wish to mine the reef for its lime, is the most magnificent and well-known example of its type. Barrier reefs are separated from the shore by wide lagoons, while the atolls grow to form horse-shoe-shaped reefs enclosing a protected lagoon.

A few of the corals are solitary and do not build reefs but generally the species are made up of colonial animals dependent upon each other for survival; both the colony and the individual are interdependent. The way in which community living can affect survival is also illustrated by many higher forms of life and often presents a baffling anomaly. For example hippopotamuses congregate in such dense numbers that they

A female orb-web spider, *Araneus cucurbitinus*,
waits for prey. This species can adapt the
construction of its web pattern to suit the
surroundings and so does not always produce a
complete and typical orb-web.

The social paper-wasps have developed a system of construction similar to the honeybees. They build a collection of hexagonal cells from wood scrapings mixed with saliva which dries to form a delicate papery substance. Rigidity is obtained from the shape of the construction rather than the material.

apparently cause nothing but discomfort and trouble to each other. But they cannot live alone and when isolated in zoos are merely existing, like a man in solitary confinement. Similarly one cannot discuss the social phenomenon of the beehive in terms of the behaviour of one bee; as individuals they lose their identity and function.

Humans also crowd into city conurbations and have to argue for space. The same stimulus is common for both social man and animal; perhaps it can be most easily explained as a need for togetherness. There is however one subtle but essential difference in these arrangements. It is a factor which allows the animals to develop successfully but is being ignored by man at his own peril. Those animals which live in such apparently over-crowded conditions never exceed a density which prevents their specific individual distances from being maintained. In some species this requirement of space around the individual is so reduced that it can be measured in millimetres. At the other end of the scale others, such as the hunting eagles, maintain hundreds of metres of space between themselves and their neighbours.

Swallows resting on telegraph wires demonstrate how accurately this spacing is measured and this same distance is maintained between their nests. In the same way gulls nest just beyond pecking distance of each other. It is this basic requirement which has been ignored in the planning of our cities. Individuals cannot create the necessary individual distance between themselves and many are deprived of their need for private space and temporary seclusion. In these conditions the potential benefits of city living are thus transformed into real hazards. But animals, when they build their social units, are not influenced by such considerations as profit.

The effect of external artificial factors upon a society are clearly illustrated by the red-billed weaverbirds. These birds live in colonies of up to ten million individuals. There are recorded instances where almost 2,000 nests have been built in one tree and branches are sometimes broken off by sheer weight of numbers. Living in such dense communities obviously brings both its benefits and its problems, as does a solitary existence. A serious danger to the weaverbirds is the constant threat of attack from eagles, herons and hornbills which are always ready to take the nestlings. The weaverbirds combat this by choosing the protection of thorn trees for their nesting sites and by packing their nests into the centre of the tree. However they are unable to deal with what is for them a relatively new problem. Native villagers, to protect their crops from the birds, persistently burn down the trees. And no degree of ingenious planning or nest-building can ever beat this artificial hazard.

Other parallels between the organisation of human society and that of animals can be found with some of the social insects, and especially the ants. Though any comparisons can be misleading, it is true that the behaviour of the ants has always attracted great attention and the example of their labour has influenced many great men. The mathematician Malchus contemplated the ants and resolved to lead a more industrious life; St Jerome studied their behaviour and learned that men should work together and share common property. It is not surprising that mankind, as Solomon dictated, should be able to learn from the ways of the ant for we, like them, live in organised communities protected and pervaded by law and order. They rely on sophisticated methods of communication, and are dedicated to hygiene for reasons of survival. They are masters of agriculture, and they harness the labour of the individual for the benefit of their society. But there are other similarities between ant and human activities which are less desirable. The ants for instance also have armies and they practise infanticide and genocide. They have slaves and they have war.

The first primitive ants appeared on the earth some 30 million years ago. Today some 4,000 different species are classified, none of them solitary, and their population probably exceeds that of any other terrestrial animal. Their communes are invariably large, successfully organised and self-supporting. The first ants, which evolved from some kind of solitary wasp, probably nested in the soil or in rotting wood and either hunted their prey on the ground or collected dead insects. Nesting in natural crevices above ground and feeding on seeds, fungi and honeydew were later developments, as were group activities such as collective hunting and building. The ants are now spread throughout the world, having colonised a wide range of different habitats, and their various activities have become characteristics of certain specific groups.

Within each species there are normally two castes, the males and females who are concerned only with reproduction, and the sterile workers. The latter have to forage for themselves, feed the queen, care for the young and build and defend the nest. Some species breed special workers equipped with large mandibles as soldiers, and many of the other workers are often grouped into specific divisions of labour. Individual wood ants sometimes tend to express an aptitude for certain activities and become so special-ised that they may spend their entire life carrying out one particular type of work. The other workers however are always trying different tasks. They may work at one job for a few days, perhaps even weeks, then try another. There are also the idlers who are either sedentary or else rush about doing nothing constructive. The incessant activity of an ant community is in fact something of a cover-up for incompetence and at any one time about half the workers are inactive. Furthermore while one ant is working as a builder it tends to ignore any other tasks or opportunities unless an emergency arises. Similarly a worker engaged on hunting will not attempt to repair any damage to the nest, not even a partially blocked entrance.

When an ant is about three months old it is usually at the age to start work outside the nest. At first it merely spends its time at the entrance to the nest helping to pass debris out to the other workers or receiving food from them to take back into the nest. Soon it will begin to venture out into the nest mound itself, assisting with repair work and removing grains of sand and rubbish. When it does begin to follow some of the scent tracks leading from the nest it will not travel any great distance but returns home by helping other workers to carry their burdens of food or building materials over the last lap. Finally, after gradually extending its range, it becomes a fully-fledged worker, foraging for pine needles and food, tending the aphids on the farm and later setting out on expeditions with other scouts to investigate the local territory and extend the empire.

The most common type of nest, and the most primitive, is built underground, and the new queen often chooses a site where there is some sort of existing roof to protect the nest, such as a flat stone or a log or, in cities, a pavement slab. The underground nests vary between the species from a maze of shallow tunnels, which may be merely small scale extensions of the original brood chamber, to deep vertical shafts containing a great number of chambers with galleries extending several hundred metres in length. The main advantage of an underground nest is the access to moisture in the soil. In dry and sandy regions deep vertical nests are common, and small brood chambers leading off the main shafts have sloping floors so that any collected water drains out into the communal tunnels. Typically, in the northern hemisphere such species as wood ants build a mound above their nest which helps to insulate the interior. Temperature control is critical for ants cannot regulate their own body heat and are acutely suscep-tible to the smallest changes in temperatures. Furthermore their choice of a select

The impressive size of a wood ants'-nest, which
can often exceed a height of 1 metre, is the result
of months of labour by millions of individual ants.
The densely thatched insulating cover of leaves,
sticks and vegetable debris is constantly
maintained in good order throughout the
summer days to ensure comfortable conditions
within the nest.

Workers of the weaving ant *Oecophylla smaragdina*. The leaf edges, having been pulled together, will be closed and held with larval silk.

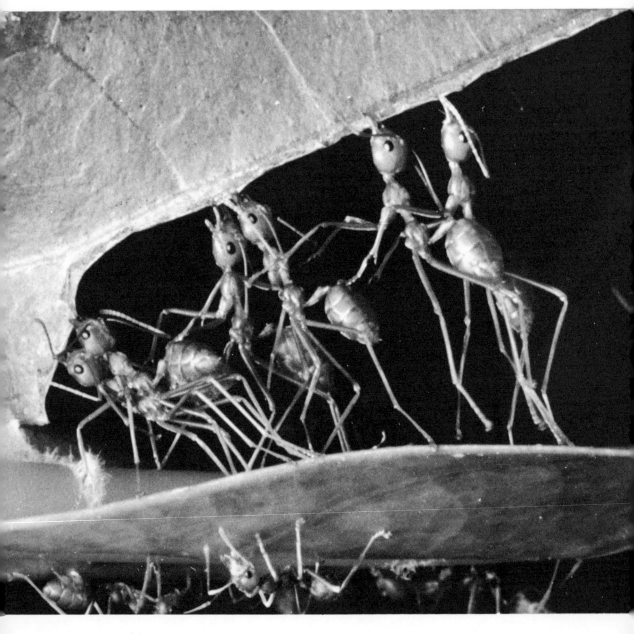

temperature varies from season to season and their movements around various parts of the nest correspond with these requirements. In summer for example the queen seeks a temperature of about 21 degrees C but in the autumn and spring she prefers a temperature of about 29 degrees C.

Although the temperature of the nest is not, as with the bees' hive, affected by the body heat of its inhabitants an inhabited nest is nonetheless much warmer than one that has been deserted. The clue to this phenomenon is found in the ants' behaviour and is one more indication of the inter-relationship between building behaviour and the natural environment. The ants are constantly manipulating the surface of the mound to control heat loss. Individual workers thrust head and thorax under a pine needle, lifting it up in much the same way that hairs on the skin are raised when one shivers. In like manner a bird fluffs out its feathers and a hamster fluffs out its newspaper nest. In addition semi-permanent holes may be formed in the top of the mound as ventilation shafts which can be enlarged or reduced according to the particular requirements of the day. On cold and misty days they remain closed as do the main entrances at the base of the nest.

During winter a mixture of fine particles of various materials soaks down into the nest, gradually clogging the spongy texture. Early each spring the workers, who have spent the winter clustered together inside the dark nest, have to clean the structure of the mound, opening up the spaces again and restoring its insulating qualities. At this time fresh building material is also gathered and added to the covering of the nest. In the summertime the surface of the mound is continually turned over and re-arranged by the ants and any foreign objects are removed. This ensures that the smooth and streamlined silhouette of the mound is maintained and is therefore less likely to be damaged by wind or rain.

By keeping the thatch in good condition the ants help to maintain equable conditions inside the nest. They are also able to affect the amount of heat gain or loss by altering the shape of the mounds. During experiments with infra-red lamps the insects formed a flat mound when the heat source was placed near to the nest but when the lamp was moved farther away they began to pile their mound into a heap. Thus in natural conditions the mounds are usually constructed with the longest slope facing south to receive the greatest amount of sunshine. Similarly ants in the high Alps often seek a location sheltered by a large stone from the east winds. It has been suggested that some worker ants sunbathe on top of the mound until they are baking hot and then dash down into the core of the nest carrying their heat with them. Whether or not the queen ant in her wisdom prepares a special batch of heat-carriers it is certainly true that the ants spend a great deal of their time and energy in trying to maintain comfortable temperature conditions in the nest.

Those ants which grow fungus foods within their nest find this problem even more critical for the fungi only grow successfully in a range of conditions so limited that man has not yet been able to grow many of them in a laboratory. Growing their own food underground allows the ants much greater independence but it is an activity which only a few species have developed and these are all to be found in the New World. In general fungus growers build deep and complex nests and maintain very large communities. Among the most well-known are the parasol ants whose name is derived from the curious manner in which they cut small circles from leaves and carry them over their heads. These small portions of leaf are taken into the nest where they are thoroughly cleaned by scraping and licking, then chewed up and taken down to the gardens to be pushed firmly into one of the vegetating fungus beds. Other species collect caterpillar droppings as a substratum for the fungus, or sawdust or insect skeletons. The gardens are constantly attended and weeded by some of the smaller weaker workers and the beds are also often manured with the ants' faeces.

Other species of ants are unable to grow their own food but some have developed ingenious methods of storage. Harvester ants collect grass seeds which they husk and store in underground chambers. Honeypot ants have evolved an extremely practical but macabre method of food storage; they use certain workers as living storage vessels. These repletes as they are called spend their life hanging upside down from the ceiling of the communal larder, receiving honeydew from the foraging workers, storing in their grossly distended crops and regurgitating it when required. They are forced to stay in this position, for if one should fall it dies, sometimes by exploding on impact. But in any case it is completely helpless and its fellow workers are unable to lift it back to its perch.

Ants are not restricted to underground living. Many species have either evolved resistance to desiccation or methods of collecting water and have been able to explore a variety of nesting habitats above the ground. Having abandoned the moisture-bearing soil these ants are continually forced to seek artificial ways to control the humidity of their nests and evaporation is often rigorously controlled. One such species found in both Europe and America breeds workers with large flat heads. Their sole function is to stand at the entrances with their heads blocking the opening, moving aside only to allow the smaller workers in and out.

Rotting wood has many obvious similarities with a soil habitat and is thus readily favoured by many ant species. The carpenter ants often excavate their nest in dying trees, following the softer growth rings as they dig down into the roots and upwards into the trunk of the tree, and hanging their larvae from the walls. The ants abandon the upper reaches of their nest in the winter, sealing it over and lagging it with layers of sawdust. The black jet ants also build their nests in rotting tree stumps, constructing the partitions with carton. This is reconstituted wood, similar to but much denser than the papier-maché made by some of the wasps. It is produced by chewing wood fibres into a pulp which is mixed with a gland secretion. The walls which they build with carton often become permeated with a thread-like fungus which reinforces the structure while the offshoots also provide food.

In south-east Asia giant egg-shaped nests can be found, up to one and a half metres in diameter, built around branches high in the trees. These are made of a mixture of carton and soil carried up on countless trips from the ground. This large and complex structure begins as one small flake attached to the bark of a tree. Other flakes are added and firmly cemented to each other until the first cells have been formed and then layer upon layer of inter-connecting cells are constructed. Finally the only flakes left as a visible reminder are formed as small porches jutting over the numerous entrance holes. Nests of similar appearance occur in parts of Africa, made of a carton which when dry is as hard as any wood. A particularly large nest of this variety can take over 20 years to complete.

A more delicate form of arboreal nest is that built by the tailor ants; it is about large enough to be cupped in the hands but is of a shape and construction which offers maximum resistance to rain, minimum resistance to wind and is strong enough to withstand hard knocks. The nest itself is remarkable but the method of building is even more so. The tailor ants use green leaves as a basis for their nest and as these die new constructions are continually undertaken. The colony therefore has small groups of scouts which set out to discover new areas suitable for colonisation. The essential requirement they seek is a place where two leaves are growing near to each other in such a way that they can be drawn together. One ant, holding on to a leaf with its hind legs, will stretch out until it can reach another leaf. If it is able to grip this leaf other ants will climb over its body and test whether or not the two leaves can be pulled together. This is all the scouts do for the time being, but presumably they report back, for later in the day other ants appear at the scene and again begin to draw the leaves towards each other. Five or six ants grip each edge of the two leaves, their bodies stretched out roughly parallel to one another. They then pull the leaves closer until more ants can join the effort and span the narrowing gap. After anything up to three hours they will have pulled the leaves far enough for the edges to touch whereupon they hold them firmly and steadily while another batch of workers appears, each one carrying an ant larva in its jaws. They now carry out an activity which is co-operation carried to an extreme, for the larvae are induced to spin their silk, which must originally have evolved for cocoon spinning, to bind the nest together. The workers carry the larvae to the leaf edges and begin to stroke them with their antennae to induce the spinning of silk. At the same time the head of the larva is placed against the edge of the leaf and moved backwards and forwards in a zig-zag sewing fashion from leaf to leaf, binding the leaves together with a sheet of continuous silk. When the construction is finished the larvae are then housed in the new nest they have helped to create.

The tailor ants' nests are far removed both in space and fashion from the underground tunnellings of the more primitive ponerine ants. And even more so from the army ants and their warlike way of life. Almost all ants are pugnacious but the army ants of South America and Africa are by far the most aggressive. They are perpetually at war; a great army of soldiers, totally blind and incessantly attacking all other animals. Because they are nomads they build no permanent nest. Indeed everything they do is achieved by the use of their own bodies. They construct living bridges to cross streams and are said to survive floods by gathering themselves into a tight ball which floats on the water.

New camps are established each day just before daybreak; the ants either seek shelter under a log or a rock or else they cluster in one large swarm with the queen and larvae protected in the centre. During their marches the ants half drag, half carry their heavy queen with them but occasionally they spend what is called a statary phase when they stay in one camp for about three weeks. Here they gather in a swarm, hooked together by their legs and enclosing the queen and her eggs. In one week she will lay well over 100,000 new eggs and a few days later the previous generation of larvae, which the ants have been carrying in their cocoons, hatch almost simultaneously. The dramatic increase in population excites the ants into another cycle of nomadism. They once again begin their series of marauding parties, carrying the new larvae with them by night and seeking shelter each day.

Termites share many superficial similarities of behaviour with the ants but they are not related. Ants belong to the order *Hymenoptera*, which includes the bees and wasps, but termites, the most ancient of all social insects, are in an order of their own and their nearest relatives are the cockroaches. Divided into six families the termites all live in colonies and have a sophisticated caste system of kings, queens, soldiers, workers and nymphs, each of which develops differently anatomically and has a specific function in the society. Their original ancestors probably developed social groups through the

The ungainly dramatic forms of such giant termitiaries are a common sight over large areas of West Africa. The mound encloses a complex labyrinth of air ducts to ventilate the nest. The palm trees are in the background.

Right Illustration from the 'Account of African Termites' given to the Royal Society in 1781 by Henry Smeathman. The drawing includes a section to show the construction of a nest of the genus *Macrotermes* which, he claimed, contained streets, bridges, canals, food stores, nurseries, guard rooms and a royal palace.

habit of burrowing in rotting timber where they would of necessity have to live in close and confined proximity. It seems likely that they then developed the habit of eating each other's faeces, for the droppings would contain undigested nutritive particles, and from this action mutual feeding and grooming would easily have evolved, helping to bind the groups into true colonies.

There is a remarkable and complex variety of nest structures among the termite families and the type of nest produced by any one species tends to be dictated not by any generic similarities between groups but by their type of food. Thus wood eaters make nests of carton while soil feeders build with their own excreta, and those termites which cultivate fungus gardens tend to build mainly with cement made of sand and saliva.

The simplest nests are found in the families *Kalotermitidae* and *Termopsidae* such as are found in the rain forest areas on the Pacific coast of the USA. These are merely random excavations in damp and rotting timber, with the confused network of galleries following the grain of the wood. This type of nest is also excavated by many of the dry-wood termites of southern Europe, but these also attack seasoned timber and can cause serious damage. Napoleon's ship *Le Génois* had to be broken up in 1820 because of the damage that had been inflicted upon its superstructure by termites. The destructive abilities of the wood-eating termites are universally well-known and it is estimated that in the USA alone they cause damage in excess of $100 million a year. Apart from wood the termites also attack plastics and metal foil. They have been known to eat through lead piping and on one occasion reduced a set of billiard balls to hollow shells.

The mound-building termites however can have substantial benefits on the land.

Some species build nests on the African savannah up to 30 metres in diameter and six metres high. These structures are rich in organic material due to the inclusion of faeces as a building material, and also have a high mineral content of clay which the termites bring up from near the underlying water table. By their construction the nest mounds are also well-drained and when they are abandoned are rapidly colonised by grasses, shrubs and then trees which eventually spread to form continuous forest growth. Termites also benefit the land by aerating and draining the soil, probably preventing erosion. The nest mounds also have some economic importance and are eaten by some primitive tribes who thus add rich amounts of mineral salts to their diet. When crushed into powder the material of the nest can be mixed with water to form a cement and in this way has been used in road construction. Some indication of the amount of material available is shown by one nest mound which was broken up and made into 450,000 bricks.

The most primitive of the termites is *Mastotermes darwiniensis* of northern Australia. Their way of life however is surprisingly sophisticated and their nests are of elaborate form. They live in trees, either living or felled, in colonies numbering millions of individuals, and excavate wide-spreading underground galleries which grope outwards from the nest in search of new sources of timber. There are two to three thousand species of termites, almost all tropical, and *Mastotermes darwiniensis* is unique in that it is the only representative of its family. There are however hundreds of other species which have adapted to living in timber and many of these belong to the family *Rhinotermitidae*. One of the European species excavates in damp wood, lining its galleries with carton made of wood pulp, faecal matter and particles of earth. With this material they are also able to build tunnels and bridges, either horizontally or vertically, so that they can bypass large areas of rock or brickwork and colonise other areas. Another species of the family, living in Australia, attacks living trees. These termites build large mounds of clay around the base of a tree, sometimes to a height of over two metres, and the nest is built inside the enveloped trunk. When the tree dies the nest expands to fill the enclosing mound. Other tropical Australian termites build mounds almost eight metres high, their bulging and grotesque forms lending a surrealistic quality to the landscape.

The remarkable compass termites of the genus *Amitermes* are also found in Australia. These belong to the largest family group, the *Termitidae*, which comprises over two-thirds of all known species. The slender tapering nest mound of the compass termites reaches a height of up to four metres and is always located on a north-south axis. In this way the minimal-surface knife edge of the mound points towards the intense heat of the noon sun and the flat broad faces of the structure are presented to the lower forces of irradiation at morning and evening.

Other *Amitermes* species store grass and seed foodstuffs in their nest and this activity foreshadows the behaviour of the fungus growers. These termites belong to the remaining family *Hodotermitidae* which includes the genus *Macrotermes*, a species of African termite which builds the largest of all nests. This nest begins as an underground chamber where the termites develop a fungus garden surrounding the royal cell. As the nest grows the termites begin to build a dome above the garden and as this increases new gardens are developed around the original chamber. Slowly the structures swell to form artificial miniature mountains, up to six metres high and 30 metres in diameter, which have large scale effects on the savannah landscape. All members of the *Hodotermitidae* family cultivate fungus gardens, but their nest forms are developed in many ways. Some are completely hidden underground and others build chimneys projecting above the soil. The presence of these towers is not completely understood for they do not open on to the fungus gardens or provide ventilation as earlier naturalists had assumed. Likewise some species excavate horizontal ducts leading from the gardens but these are also blind-ended and do not give access to the outside air.

The termites which have developed the ability to grow fungus foods have become the largest and most complex groups. They are independent to a large extent of external conditions but like domesticated man have inherited other problems. To maintain both their vital food supply and their large populations they have to deal with internal difficulties such as the regulation of a favourable air temperature and purification of the air. Humidity control inside the nest is also critical, especially for those in desert areas, and in any case needs to be kept at a very high rate, well over 90 per cent. Some species in the Sahara survive by digging down to the water table, sometimes 40 metres below. This allows moisture to evaporate into the nest, and the termites also carry damp particles of clay from their mines into the interior of the nest.

The construction of the nest walls assists this control of humidity, especially with mound nests where the dense material of the structure prevents evaporation; also it is covered with a cemented layer of sand and clay impervious to moisture. Water vapour in the nest tends to be trapped by the intricate layout of the galleries. In some instances the walls of the nest are constructed in such a way that favourable temperatures and humidity are guaranteed. The domed nest of one Asiatic species, built typically of extremely thick walls, is comprised entirely of earthen material at the surface but this slowly changes in content to a mixture of earth and excrement until the inner structure of the wall is almost entirely composed of faecal pellets. This inner zone of the wall, containing the higher proportion of organic material, absorbs the greatest amount of moisture and this gradually decreases through to the outer layers of the wall. The unequal absorption rate of the nest structure ensures stable conditions at the centre and prevents the moisture from accumulating in the outer layers of the wall where it could evaporate.

Whether the nest is inside a tree trunk, underground or inside the thick walls of a nest mound the termites are always insulated from fluctuating external temperatures. But apart from protecting themselves against the outside heat and cold termites also have to disperse the accumulated body heat of the colony as well as the carbon dioxide given off by the fungus gardens. Occasionally ventilation holes are opened in the nest walls to allow the gases to escape and some nests contain an elaborate system of air ducts. These start as fairly large openings at the top of the nest, directly above the nest chamber, and subdivide into smaller branches as they reach the outer walls before joining again and running back down to the base of the nest chamber. As the capillary ducts run close to the external walls carbon dioxide is diffused and oxygen absorbed so that the nest structure acts as a lung. While passing through the duct system the air cools and falls back down the return branches so that a constant supply of cooled air, rich in oxygen, reaches the inner chambers at a flow rate of approximately 12 centimetres a minute. In this way the nest is kept at a steady temperature of about 30 degrees C and this varies on average by less than half a degree throughout the year.

The termites are also acutely concerned with hygiene in their crowded nests and are assisted by a host of scavenging parasitic mites. Dead termites are eaten by the workers and partially digested excreta is passed from individual to individual until every digestible particle has been disposed of. All that remains is incorporated into the building materials and becomes part of the walls of the nest structure. Some wood-eating species

The most ordered constructions built by termites are found in the works of the genus *Apicotermes*. This subterranean nest of *A. desneuxi*, a forest-dwelling species, shows the symmetry of the parallel storeys in section. Parts of the spiral ramps connecting the floors can also be seen, and the exit tube at the bottom of the nest. The surface of the nest is coated with a very finely granulated sandy material and is about 10 cm in diameter.

excavate cul-de-sacs where they dump the faecal pellets, although one species makes openings in the nest walls so that the faeces can be thrown outside. These openings are kept closed by soldiers who block the exit with their large heads, and they may also be sealed off temporarily with partitions of carton.

Apart from building huge mound nests which dominate the landscape African species also build extremely delicate structures. Some build on the bark of trees and protect the nest from rains with a number of overhanging canopies. Similarly small bollard-shaped nests projecting from the floor of the jungle have umbrella-like structures built over them to shed the rain. Underground nests can also on occasions be extremely regular and decorative, such as that of *Apicotermes gurgulifex*, found in the Congo. Constructed about 500 millimetres below the surface the nest structure resembles a huge pine cone and is as regular in pattern as a honeycomb.

Diagrammatic sections of the nest wall of
different species of *Apicotermes*: 1 *A. occultus*,
2 *A. kisantuensis*, 3 *A. lamani*, 4 *A. desneuxi*,
5 *A. gurgulifex*.
These various types of openings in the nest wall
are unique to each species. They are too small to
allow the passage of termites and their probable
function is to allow the exchange of gases, so that
carbon dioxide can be given off and oxygen
taken in.

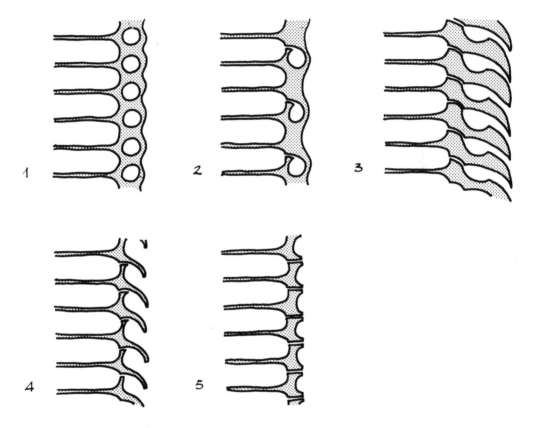

The more advanced species of termites build the most symmetrical and ordered constructions, and the genus *Apicotermes* always provides the finest examples. Some are
quite effectively simple, consisting only of a series of spiral chambers wound around
each other. Others build nests made up of horizontal storeys separated by floors one
millimetre thick. Access between the floors is by ramps, either straight or spiralling,
guarded at each level by soldiers to protect the nest from insect invaders. Ledges or
tubes projecting from the top of the nest serve as exit ramps, but leading off from each
floor is a series of small apertures in the nest wall which at first sight appear to be
access openings. However these are too small for the termites to pass through and their
principal function seems to be to allow the diffusion between carbon dioxide inside the
nest and oxygen in the soil. The apertures, which are either in the shape of small slits
or circular holes, are always protected by some kind of overhanging structure; these may

The subterranean nest of *Apicotermes gurgulifex*, a termite species of the African savannah. Part of an exit tube is visible at the top, and the pattern of the gargoyles with their slit openings are clearly seen. Konrad Lorenz has appropriately termed the *Apicotermes* nests 'frozen behaviour'.

The pattern of construction of a *Macrotermes* termite mound controls the temperature and oxygen supply in the nest by a system of convection currents. The activity of thousands of termites in the nest warms the air and increases the carbon dioxide content. This heated air rises into the hollow chamber above where a number of radiating tunnels, each about the thickness of a man's arm, direct it to the outer walls of the mound to flow into a system of vertical capillary ducts. The air loses some of its heat, and absorbs fresh oxygen from the outside air as it falls. At about ground level the capillary ducts open up to form branches about 10 cm wide; these lead into the nursery area in the cellar of the mound.

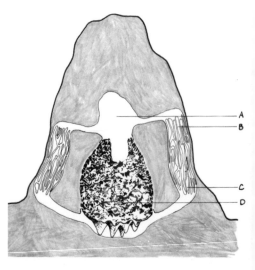

Microclimatic measurements:

	Temperature (degrees C)	Carbon dioxide percentage
A	29	3·0
B	25	2·7
C	24	0·8
D	30	2·7

be to prevent water seeping into the nest. The surprisingly different types of construction of these slits and overhangs vary between the species of *Apicotermes*, each displaying the instinctive habits of the termites in what Konrad Lorenz has termed 'frozen behaviour'.

The walls around the horizontal galleries in the nest of *Apicotermes lamani* have a distinctive series of slit openings which are bottle-shaped in section, with the elongated neck of the bottle leading off from each floor. *Apicotermes gurgulifex* on the other hand has developed the slits into rows of modernistic gargoyles, reminiscent of Le Corbusier's work, whereas *Apicotermes desneuxi*, while following a similar principle to this, builds projecting ledges above each slit. Occasionally these curve over to such an extent that they fuse with the next canopy below to form short sections of horizontal tubes. This pattern of construction is carried to its logical conclusion in *Apicotermes kisantuensis*, where the tubes have coalesced as it were to form complete rings of galleries around the nest at each floor level. The slit openings from each storey curve open into these galleries and the outer wall is then perforated with small circular openings built in an alternating pattern with the slits.

It is even more difficult to conceive *how* termites build than it is to explain the structure of their nests. But it seems likely that construction is carried out by a series of 'releasers', in much the same way that birds build their nests. They seem to possess an insatiable urge to build and this habit is probably prompted by changes in external conditions. At first they build only at random but as some sort of exploratory structure evolves from this activity it acts as a stimulus. The particles of building material are first built up to form pillars which then stimulate a change in the design pattern so that the termites begin to build horizontally. Eventually arches are constructed between the pillars and these are extended to form a roof. The air duct system develops as the nest structure is built up, due to the extreme sensitivity of the termites to air movements. Convection currents set up by the shape of the developing nest act as an invisible scaffold and the termites build around these air movements, preserving them in a solid casing. The termites, though blind, are also sensitive to light, building to seal it out and plugging any points where light or draughts may enter. In this way, by reacting to specific stimuli, each stage of the construction progresses in a programmed and predetermined sequence.

For elegance and precision, for structural ingenuity and for powers of improvisation the animal builders are, as Sir Hugh Casson has pointed out, better at architecture than we are. And perhaps none are more competent than the termites. Their structures, ranging from delicate and fragile arboreal nests to wide-ranging underground caverns and huge monolithic towers, are rarely matched and it is understandable that Henry Smeathman when giving his account of termites to the Royal Society in 1781 should have exclaimed that they 'appear foremost on the list of the wonders of the creation'. But the success of the termites should not overshadow the wonderful achievements of the other animals, and the nests of a tailorbird, the webs of the spider and even the slender tubes of the fan worms are equally magical and impressive constructions.

The building abilities of all social insects have evolved to become articulate expressions of their ordered way of life. The social wasps for example do not build such permanent constructions as the termites, for the colonies do not survive the winter, but their honeycombed nest is nonetheless an efficiently built and well-arranged structure. The nest lasts only for the summer, for only the young fertilised females have the ability to hibernate until the spring when they begin to establish a new colony and build a new nest.

The common European wasps build underground, often in disused mouse holes, but a few other species nest in bushes or inside lofts. Using her powerful jaws the young queen wasp shaves scrapings of wood from posts and dead trees, mixing it into a pulp with her saliva. With this material she builds the first few brittle cells for her eggs, constructing a nest about the size of a golf ball. After the first workers have emerged they take over the task of building new cells, extending the structure into horizontal layers of combs. The colony expands throughout the summer and more workers are born to collect food, care for the eggs, feed the larvae and build more new cells. At the end of the summer a large and successful nest may contain some 2,000 individuals, including a few males and females which in the early autumn leave the nest for their nuptial flights. The new queens then search for a winter shelter to hibernate, continuing the cycle and carrying the species through the winter.

Hornets, the largest of the European wasps, build very similar nests, except that they are found only in wooded areas, normally nesting in hollow trees. The other main difference is that hornets, though much maligned and persecuted, are probably the most docile of all the social wasps and are much rarer than the more pugnacious but smaller common wasps.

Bees also often receive harsh treatment, and perhaps some find justification for killing bees and wasps indiscriminately when statistics reveal for example that 229 people were killed by bee and wasp stings between the years 1959–63 in the USA. But even without man the bees have a staggeringly large number of natural predators. Bears can be a serious menace to commercial apiaries in parts of North America, and termites eat away the wooden structure of domestic hives. In hot climates ants often kill the bees and steal their honey; robbing wasps make mass attacks on beehives and digger wasps capture bees to feed to their larvae. Spiders sometimes spin webs over the hive entrance, and toads and lizards also station themselves there, noisily snapping up the bees as they emerge. Dragonflies and some birds also take bees on the wing, especially the African bee-eater. In European winters the tits even tap on the hive to excite the attention of the bees, snatching away any individuals who come outside to investigate, and skunks employ a similar practice, scratching on the hive to attract the inhabitants. Woodpeckers break into the hives to reach the bees and in winter, when the bees are not likely to attack them, mice sometimes enter and become so gorged on honey that they cannot escape through the flight hole.

A great deal of this unwelcome attention is because bees are able to produce and store large amounts of honey. Since man has learned to domesticate the bees and control their activities to suit his own requirements he no longer has to rob the hives and has given more of his attention to studying the social behaviour of the bees and their methods of building and production. The honeybee probably originated in south Asia and spread out to China, Europe and Africa. They were introduced to the New World around 1530 when the Portuguese took them to Brazil, and in 1638 the Dutch established honeybees in North America. Bees then set out to colonise the continent, spreading westwards long before the first pioneer settlers. In the wild they seek shelter in a natural cavity, such as a hollow tree, and there make their nest of hanging combs, some to contain honey and pollen, others for the unfledged larvae. Most of the 20,000–50,000 bees in such a hive will be workers who are equipped for a variety of tasks; they have glands for producing bee-milk to feed the queen and the larvae; other glands for converting honey into wax for building the cells; and a third set of glands for converting nectar and pollen into honey.

Ammophila-sphex, a solitary hunting wasp,
excavates a burrow and furnishes it with a
paralysed caterpillar on which she lays her eggs.
She then fills in the hole with loose earth which
she tamps down with a pebble held in her
mandibles.

The male bower-birds build elaborate
constructions of twigs and vegetation where they
woo their potential mates. To compensate for their
lack of brilliant plumage they decorate these
bowers with bright or glittering objects. This
greater bower-bird has added various white stones
and shells, bottle tops, petals, pieces of broken
glass and bits of red plastic for decoration.

There is a distinct and functional arrangement in the way the combs are utilised. The brood for example is kept in one special and compact three-dimensional area, which eases the problems of nursing and temperature control. Adjacent to the brood nest the pollen is stored, for the nurse bees have to feed on this to produce the vitamin-rich bee-milk. Beyond the pollen storage area the bees store their honey. Any unnecessary gaps which may remain after the combs are built are plugged with propolis, and some species from eastern Europe also use propolis to build curtains across the entrance for winter. Propolis is gathered as the resin which bleeds from many trees, and among its other uses it is also an antibiotic, helping to control the growth of any mould in the hive. Furthermore if any small mammal such as a mouse should be attacked inside the hive and killed the bees coat its body with layers of propolis to prevent its putrefaction.

The orderly planning of the honeycombs and the functional uses of the various materials within the hive are paralleled by the regulated division of labour between the bees. Guards are sited near the entrance and warn the others of danger by a distinct buzzing of the wings. Worker bees forage abroad bringing pollen, nectar, water and propolis back to the hive. The house bees pack the pollen and nectar into the cells, sealing them over with wax. They also carry out the critical functions of temperature and humidity control. Operating in teams distributed around the hive they fan their wings to divert air currents in the most advantageous directions. The temperature in the hive can fall below 15 degrees C and the bees will remain active, but the optimum temperature required at the brood nest must not fluctuate more than one or two degrees from 35 degrees C and this has to be maintained irrespective of external conditions. Extreme heat is particularly distressing to the bees and in hot weather hundreds of workers are co-opted as temporary house bees driving air currents through the hive with furiously buzzing wings. Humidity is also controlled within the hive and in the brood area is maintained between 35 and 45 per cent. Excess water vapour tends to be removed by normal ventilation but in hot, dry conditions full air-conditioning principles have to be put into operation. The workers then bring water back to the hive in their crops and droplets are smeared over the combs. As this evaporates it helps to chill the air and increases the general water vapour content.

The house bees also have the task of cleaning the hive. All litter and debris are removed along with dead workers and other insects. They also have to clean out each cell before it is visited by the queen. She then inspects it, rejecting those which are unsuitable, before laying an egg inside. The number of cells which have to be built is indicated by the fact that this process of egg-laying continues like a machine operation throughout spring and summer. At peak rate the queen is laying 2,000 eggs a day. The egg laying is carried out in a methodical arrangement, the queen working concentrically from a central area. This ensures that the nurse bees are able to care for groups all of the same age in any one area.

Cell construction is a social phenomenon and is dependent upon some means of communication between the queen and all other members of the colony. No individual house bee ever carries out a complete and uninterrupted sequence of cell construction. She may begin a new cell with wax from her own glands then move to work on another cell which is in the final stages of construction, using perhaps wax taken from an old cell no longer in use. Following this she may move on to feeding the larvae or attend the queen and then spend some time planing and polishing the inside of a new cell. Thus each individual may prove herself capable of carrying out each action in building a cell but will never build one complete cell unit. This is because the programme of

Far left A female paper-wasp feeding nectar to one of her larvae in the nest she has constructed from wood scrapings mixed with saliva to form a pulp. Developing larvae (*top left*) have spun cocoons over their cell and after pupation will emerge as adult wasps.

Many of the works of insects appear too regular and naturally formed to be artificial constructions, and the hornets' nest is an excellent example. John Josselyn, exploring North America in 1638, 'chanc't to spye a Fruit as I thought like a pine Apple' only to find, too late, that it contained hornets which severely attacked him.

work is determined not by the bee's physiological state but by the changing requirements of the colony as a whole. In the same way that the termites build their nest with a programmed series of 'releasers', so also the colony of the honeybees is regulated and the beehive operates as a single organism rather than a collection of independent insects.

The first step in constructing new combs for storing the eggs and honey involves setting out a string of wax on the ceiling as a guideline. This is then extended as a midrib and the cells added to each side as work progresses. Each cell has a slight upward slope to enable it to hold the liquid honey and the mouth of the cell is slightly thickened to give strength to the rim of the structure. The shape of the cell is determined by the fact that in any collection of equal-size circles each one is in contact with six of its neighbours, and when any lateral pressure is applied to such a collection, either from within the circles or without, they will be naturally converted into regular hexagons. Thus the bees' cells begin as a layer of cylinders which become pressed together and deform into hexagonal prisms. Not surprisingly the architecture of the bees' honeycombs has attracted excited attention throughout history. Pappus, an Alexandrine, recorded his conclusion, based on the theorem of Zenodorus, that of 'the three figures which of themselves can fill up a space around a point, a triangle, a square and a hexagon, the bees have wisely chosen for their structure that which contains the most angles, suspecting indeed that it could hold more honey than either of the other two'. The regularity of the hexagonal comb is remarkable (the angles for example are constant to a limit of less than 4 degrees) and because they vary so little in size, Thévenot, a colleague of Swammerdam's, suggested in the seventeenth century that they be used as a basis for a system of standard measurement. If his idea had been adopted we would today have an international module equivalent to 5·37 millimetres.

The bees construct the combs in vertical formation and the cells are formed in a

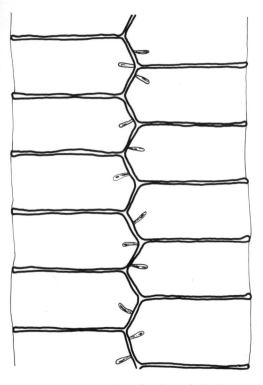

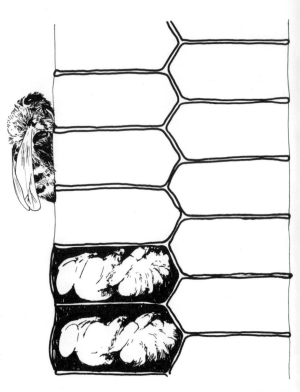

A section through the bees' comb showing the cells set staggered on either face of the midrib. The wax combs hang from the ceiling of the nest cavity, and are attached to the sides of the cavity in places to increase stability. Each of the cells on the left contains a newly laid egg, while on the right are unpigmented pupae sealed in their cells and a recently emerged worker bee. The stage of development from egg to bee takes three weeks.

Opposite Wild honeybees usually nest in sheltered places such as a hollow tree, and are thus able to maintain a permanent colony, storing food for the winter. The combs hang vertically in double layers with the cells, placed back to back, opening to the side but slightly tilted upwards.

double layer. They are not however placed directly opposite each other and the base of any one cell in the formation is in contact with three other cell bases. Just as the bees build roughly circular cells which become distorted to form hexagons so the base of each cell begins as a hemispherical basin but becomes compressed by opposing forces and is fashioned into a shallow pyramid. Thus the base appears to be made of three lozenge-shaped plates of wax and the outer edges of each of these plates coincides with the six vertical walls of the cell. Fascinated by the ingenuity and precision of this construction the astronomer Miraldi attempted in 1712 to measure them geometrically and claimed to have ascertained that the large angles measured 109°28' and the smaller 70°32'. Réaumur accepted these figures and decided that this mathematical pattern of structure must have resulted from the need to devise a system which could hold the most honey for the least wax. Certainly it would seem to be a sensible arrangement, for

wax production is extremely expensive for the bees. Without giving any clue to his thesis Réaumur therefore asked Koenig, a young Swiss mathematician, to investigate the angles required to form a hexagonal cell in the most economical fashion. Koenig calculated these to be 109°26′ and 70°34′, almost precisely agreeing with the measurements made by Miraldi.

Mathematicians were naturally delighted. It proved how practical science could be aided by theoretical knowledge, and the construction of the bee cell became a famous illustration of the economy of nature. The fact that bees had solved a problem which had been beyond the powers of mathematicians before Newton forced Fontenelle, the *Secrétaire Perpétuel*, to pronounce that bees could not be credited with intelligence but that they used the highest mathematics only by divine guidance and command.

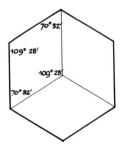

The angles of a bee cell supposedly measured by Miraldi.

Later when it was discovered that Koenig had made a slight mistake in his calculations due to inaccurate logarithmic tables and that the true theoretical angles were indeed 109°28′ and 70°32′, corresponding exactly with Miraldi's precise measurements, it was gleefully expounded that the bees were right and the mathematicians had been proved to be wrong.

It is not clear why Réaumur should have perpetuated and encouraged this theory, for it must have been clear to him as a diligent and careful observer that the bees' cells are far from being identical. They are slightly, but plainly, variable and irregular, and indeed it is obvious that Miraldi could never have measured the angles to within one minute of a degree. Furthermore it can be seen that the cells do not have a uniform thickness, the edges are never exactly straight and the angles are not geometrically sharp. In fact the pattern of the construction is determined by the laws of physical forces not by any intelligence of the bees. Although Darwin was over-enthusiastic in claiming that the honeycomb is 'absolutely perfect in economising labour and wax' this is apparently just short enough of the mark to raise scorn, but only from those wishing to apply nothing more than mathematical reasoning as a test of the bees' building capabilities.

CHAPTER EIGHT

Parasites and Partnerships

The states of co-operation and competition often overlap and in the intense struggle for survival many unrelated animals have developed associations which offer some mutual benefit. One clear and excellent example of this is found in the barber fishes which eat fungi and lice from the bodies of other fishes. The barbers thus remain well fed, and their hosts, which solicit and encourage their attentions, are kept free of body parasites and disease. In similar manner the marine iguanas allow rock crabs to crawl all over their bodies to clean them of ticks, and antelopes and rhinoceroses stand still while they are deloused by ox-peckers.

Often the partnerships which have evolved in the animal world seem bizarre and puzzling, especially those which apparently give benefit to only one of the participants, such as the example of the damsel fish living in the shelter of the sea anemone's poisonous tentacles. In other instances there are definite partnerships which seem to have no value at all. Some species of hermit crabs either place or induce sea anemones on to the shells they have commandeered. This is such a widespread and commonplace habit that it is clearly of benefit to both animals. But what the advantages are has never been discovered. Certainly with one or two species of hermits, known as grenadier crabs, the stinging tentacles of the anemones are used for protection; the crabs grip the stalk of the anemones with their claws and thrust them in the face of any attackers.

Hermit crabs occasionally seek shelter within sponges. The young crabs select small discarded shells to live in which have sponges growing on them. As the sponge grows it eventually encloses the shell and the crab. Instead of having to find larger shells as its own body increases in size, the crab carves out a spiral cavity inside the sponge. Similarly the sponge crabs cut pieces of sponge exactly the same size as their shell and with their claws transplant them on to their backs as a coat of camouflage. If sponges are not available the crab will cut and fit pieces of seaweed or even bits of rag and cardboard.

They appear to have an irresistible urge to cover their shells with something, just as the land hermit crab if unable to find a suitable shell will use the broken shell of a coconut as a temporary armour. Those types of partnership where one animal obtains shelter and protection from another, but without causing damage, are perhaps on the borderline between parasitism and symbiosis. There is another strange form of partnership known as inquilism, in which one animal lives in the home of another, sharing its food. One of the marine worms living on the west coast of North America excavates a U-shaped burrow for itself and attracts so many other residents that it is known as the innkeeper worm, although it will only tolerate certain species as guests. Goby fish regularly seek shelter in the neck of the burrow, while farther down there is often a scale worm living alongside the host and feeding off the particles of food which the innkeeper filters into its slime net. Behind the worm live one or two pea crabs and occasionally a small clam, all deriving food and oxygen from the stream of water which the innkeeper channels through its burrow.

However, inquilism is more commonly found among insects, especially the social ants and termites. Secure within the social order and thick walls of the termitiary the soft-bodied termites provide shelter for many other animals. Some 300 different species of beetles are known to take up residence in termites' nests as well as numerous mites which act as scavengers, helping to keep the nest clean. Many species of birds, including parrots and kingfishers, often carve out nesting holes in the walls, the termites sealing off the exposed tunnels but not attacking the birds in any way. Monitor lizards and several species of snakes habitually lay their eggs in the termites' nests. Even apart from

Parasitism determines the survival of all living things. Sometimes in an especially bizarre form: these bitterlings have chosen a live mussel as a spawning ground and the female (*right*) is about to lay her eggs in the animal's exhalant siphon.

such alien guests or invaders there are certain species of termites which live inside the nests of different species and feed off their supply of food.

This form of parasitism is also found among ants. One of the common North American species, *Myrmica canadensis*, tolerates and apparently encourages the impudent attentions of a much smaller species, *Leptothorax emersoni*. The latter are a rare species but whenever they are found they are always living in the underground nests excavated by *M. canadensis*. They set up residence in narrow galleries excavated in the walls of the main nest. The larger ants cannot enter but never seem to show any animosity to their small lodgers and even appear to treat them as pets, giving them rides on their backs and handing out food to them. All they seem to receive in return is a great deal of attention from their guests, which climb all over them, licking and stroking and of course begging for more food. Appealing as this scene may be the situation is in danger of evolving into a state of true parasitism, with *L. emersoni* unable to live independently.

Many other ant species have already arrived at this state and another family of the American ants, *Solenopsis*, has become specialised as a species of professional thieves. They develop their colonies inside the nests of larger species, living in small galleries like the *Leptothorax* but sneaking out on furtive sorties to steal not only food from the other ants but also their eggs and larvae. The host ants are much too large to enter the thieves' galleries, neither can they fight them effectively for in comparison they are too cumbersome and slow-footed. All they can attempt to do is seal them off in their smaller galleries but this at best is only a temporary measure.

The ant communities are plagued by over 3,000 species of parasites, including mites, woodlice, beetles and springtails. Indeed it would be unusual if other animals had not learned of the ants' food stores and taken advantage of their labours and the protection of their underground nests. Perhaps the most remarkable association is that which exists between the British large blue butterfly and two species of red ants, the common and the elbowed red ants. The caterpillar of the large blue feeds on the wild thyme plant where it hatches until it reaches a length of about one centimetre. At this stage it falls to the ground and changes its feeding habits from vegetarian to carnivorous, searching for insects to eat. But if it stays on the ground it will eventually die or be eaten unless it is discovered by a red ant. If this should happen it is carried away to live in the ants' nest where it lives a life completely different from its previous existence. It crawls into the interior of the nest until it discovers the brood chamber. Here it feeds off the ant larvae, but in return the ants receive a sweet secretion which they milk from the caterpillar. This rich and sticky fluid is a luxury to the ants for which they are prepared to sacrifice the entire colony. But they do not need this honey-like secretion, not even for emergency purposes, since the caterpillar hibernates during the winter when food supplies in the nest are low. Nonetheless the ants carry the caterpillar down into the depths where it spins its cocoon and emerges next spring as an adult butterfly and leaves the nest.

Collecting honey is an ancient habit among ants, not only from such animals as the large blue caterpillar and various species of beetles, but also from flowers, in the form of nectar, and especially from aphids. These insects tap the main food stream of the plant and their secretions thus contain many amino-acids and carbohydrates which can form a complete diet for the ants. The symbiotic existence which has developed between ants and aphids has reached its logical conclusion with one tropical species of ant which is totally dependent on one specific type of aphis which significantly is found nowhere

Man has attracted many species of animals to his home by inadvertently offering food, warmth and shelter; even an empty bottle can be put to good use by house mice.

Many species of animals have learned to fear man, and retreat (or are driven) from his presence. A few notable examples, however, have developed close associations with human habitations which moreover are encouraged. Storks commonly build on rooftops throughout many areas of Europe and Asia and, being fortunate enough to be a part of many fertility myths, are thus protected.

but in that ants' nest. Both ants and aphids benefit from this relationship and the honey-dew can perhaps be regarded as rent, or perhaps protection money, for the presence of the ants drives away the aphids' natural predators. Experiments have also shown that the aphids develop and multiply more rapidly when tended by ants.

Some ants build the equivalent of cattle sheds for their captive insects. Cement-like structures composed of earth and saliva are normally built, although weaver ants spin tent-like structures for housing their herds. These shelters are to protect them from both predators and climate. This is essential, for even a light wind will stop the aphids from feeding, thereby reducing their honeydew output. For this reason one species of tropical ants only builds shelters during the rainy season. Some of the wood ants keep their aphis herds in underground chambers where they feed off the juices sucked from roots. In this way a colony of wood ants may derive more than half of their food supply from aphids, while the yellow meadow ants live almost entirely off the products of their aphis herds which they keep inside their nest.

It is rather sobering to realise that the only other animals which ants have been able to exploit on such a scale as the aphids are ourselves. The technologies which we have developed have enabled us to colonise all types of climate, but as man has spread around the world he has taken a large number of other species with him. One of the most common of these is pharoah's ant, a former inhabitant of Egypt which now has world-wide distribution. Having taken over a house, and they seem invariably to colonise homes in middle and upper class districts because no doubt they are consistently warmer throughout the year, the ants settle into permanent residence. They establish series of colonies under floors, behind skirtings and in wall cavities. Nothing but demolition could ever remove them, though it is debatable whether these scavengers might not in fact do more good than harm.

The manner in which these ants spread, living off waste products, is dramatically illustrated by one species, *Iridomyrmex humilus*, an ant which will eat almost anything edible. It was first discovered in 1868 near Buenos Aires. Some years later it was beginning to appear along travel routes throughout South America and after a few more international trade agreements was invading Spain and Portugal and learning to hibernate in the USA. Its human-assisted migrations have taken it to Africa and Australia. It has now appeared in homes in northern Europe.

Many other small animals have adapted themselves to the comforts of our artificial environment, in particular the house flies, cockroaches, mice and rats. Invariably we seek to destroy all these invaders, but W. H. Hudson recounts an engaging story of a wild rat which befriended a cottage household in Cornwall. The lonely and childless wife of the cottager encouraged the occasional visits of the rat by feeding it and soon it became friendly and familiar. Even more strangely the cat of the house accepted this alien intruder as a privileged guest and the two animals shared milk from the same saucer and huddled together for afternoon siestas in the hearth.

The rat had a mate, somewhere unseen, and for her nesting place she chose a corner of the kitchen cupboard. She now spent a great deal of time gathering bits of straw, feathers and string and stealing snips of cotton and wool threads from the sewing basket. Unfortunately her diligence was also her undoing for she suddenly took a fancy to the soft tufts of the cat's fur to line her nest. Her persistent attempts to pluck out the hairs from the cat finally tested its temper to the extreme, and after one particularly rough attempt the cat set upon the impudent creature with lightning blows and un-sheathed claws. The rat, shrieking, fled from the house, never to be seen again.

Left The piratic flycatchers of tropical South America capture the nests of other birds. They select their prospective victims with deliberation, and spend days watching the construction of their future nursery with keen interest. When the original builders have finished their work the piratic flycatchers begin their aggressive harassment, constantly persecuting the occupants of the nests until there is opportunity to steal in and remove the eggs. At this point the fight is usually won.

Right A cuckoo chick, *Cuculus pallidei*, being fed by the host parent, a flycatcher, *Rhipidura leucophrys*. The parasitic cuckoo avoids having to care for its young, or build a nest, though the accommodation acquired for the chick is not always the ideal size.

It is common for many species of birds to nest in or around human habitations. Some, such as the house martin and the barn owl, get their name from this persistent habit. Jackdaws, as in *The Jackdaw of Rheims*, are supposed to seek sanctuary in churches, while rock warblers and swallows often nest in deserted mine shafts or chimneys. On occasions such associations can bring unexpected benefits. House martins, building under the eaves of a house, once constructed one of the safest nests ever built, having inadvertently used wet cement which had been mixed in the garden. Perhaps the most familiar birds associated with nesting on human habitations are the storks, although sparrows and pigeons have now become integral factors of the urban scene in many parts of the world. Many birds such as starlings and blackbirds are attracted into towns and villages because of the waste food we scatter about. For this reason foxes are moving into the cities; in just the same manner neolithic man first domesticated the dog. There are also many examples of animals taking advantage of the behaviour of others. In the African forests the elephants act as breakers for trails which are then made use of by buffalo, hippopotamus and eventually many smaller animals.

The animals which build nests are sometimes exploited in a more direct way. The most common example of such behaviour is that of the parasitic cuckoos, but birds' nests are also favoured by other species. The fat dormouse, unlike its more common and vegetarian cousin, takes the eggs or nestlings of birds and then takes over the empty nest. Pine martens often utilise a crow's nest or squirrel's drey, and stoats and weasels sometimes make use of rabbit burrows. On one occasion an occupied hawk's nest was commandeered by a cat which there gave birth to three kittens. Similarly the house sparrows regularly evict house martins from their recently completed nests. But a more legitimate method of obtaining a home without having to build is to take over a deserted nest or burrow. The green sandpiper makes use of old nests in this way, usually

taking over those built by song thrushes, and some of the mina birds habitually nest in the vizcacha's burrows which when abandoned are taken over in succession by swallows.

Loose associations are sometimes formed when different species share nesting sites, usually for protection. Thus waxbills and sunbirds will build their nests in close proximity to wasps or other aggressive social insects. Although other intruders to the area will be attacked by the wasps they appear to ignore the comings and goings of the nesting birds. In the same way many small birds' nests are built beneath or near to that of a bird of prey, for raptorial birds, like most carnivorous mammals, will not kill on their own doorstep. Weaver birds are therefore often found nesting in the same tree as a black kite, and wagtails with ospreys, while in the Arctic snowy owls or falcons often share a nesting site with geese. The only known regular association between a bird and reptile is that of the New Zealand tuatara lizard which lives in the shearwater's burrows, although some lizards and geckos use woodpeckers' holes for temporary shelter.

The safety of the subterranean towns excavated by the prairie dogs attracts other animals seeking shelter on the flat plains. Burrowing owls, typical of their group, rarely make their own nest and are often found in the prairie dog's tunnels, and rattlesnakes regularly shelter in the burrows. It may be that these inhabitants are equally attracted by the possibility of capturing young prairie dogs, but a variety of other species also make use of the burrows either as a winter retreat or for daytime shelter. These include such diverse animals as tortoises, bees, collared lizards, racoons and even skunks. Similarly badgers occasionally share their dens with foxes although sometimes these drive away the original badger inhabitants.

When associations such as these cause inconvenience to one of the participants the situation has developed into parasitism and in extreme conditions this leads to destruction of the host. Such behaviour exists in many guises and one of the typical examples is that of the ichneumon flies. They belong to the insect order *Hymenoptera* which include the bees and wasps, but unlike them the ichneumons have not evolved any methods of building durable shelter for their offspring. Neither is their work productive, for they lay their eggs in or on other insect hosts on which their larvae feed. Many of the ichneumons have a long tough ovipositor with which they can drill deeply into wood to lay their eggs in the body of burrowing grubs. The body of the host grub then provides the larvae with both food and shelter, the grub remaining alive and continuing to feed until it is consumed.

The parasitic wasps are closely related to the ichneumons, but some of these are

Resembling hummingbirds, the jacamars of South and Central America nest in burrows, loosening the soil with their long, thin bills and kicking it away with their feet. The nests are sometimes dug in the ground or in steep banks, or sometimes, as shown here, in the mound of a termite nest. Strangely the termites seem to tolerate this invasion and simply seal off their nest where it has been broken into.

Small birds will on occasions nest beneath or very near to a bird of prey and, in this close proximity, are safe from attack.

parasitic upon plants rather than on other animals. These are the gall wasps which lay their eggs in the living tissue of a plant. The presence of the developing larva induces the host plant to form a gall in which the larva then lives safe and well-fed. The galls are commonly found on young oaks and rose bushes, but each different species of larva induces the plant to form a gall of characteristic shape and colour. They include a variety of forms: some are spherical like hard cherries, others flat and spangle-shaped. There are the large and spongy oak-apples, the clustered currant-galls and the bedeguar galls formed of a scarlet mass of filaments like a pin cushion. The galls of *Neuroterus saltatarius*, found in California, fall from the plant and the actions of the larva inside cause them to jump. Similarly the larva of *Heterarthrus aceris* lives inside sycamore leaves, separating the upper and lower skins of the leaf to form a protective blister. When developed the larva spins a spherical cocoon about itself, attached to the leaf, and this falls to the ground where by flexing its body in alternating rhythms the larva is able to make its cell jump and so finds its way into a suitable niche for hibernating.

Of those wasps which are parasitic on animals one of the species, *Melthoca ichneumonides*, practises a particularly hazardous and cunning form of parasitism. It lays its eggs in the body of the tiger-beetle larva, which is a rapacious carnivore and is equipped with large and vicious mandibles. This larva digs itself a trap about one foot in depth and waits concealed in the neck of its burrow for an insect to approach. With its hooked mandibles it seizes its prey and drags it down to the bottom of its burrow to be eaten alive. The wasp however has learned to take advantage even of such a dangerous ambush. Indeed it seeks out the lair of the tiger-beetle grub and allows itself to be captured. But at the foot of the burrow the wasp suddenly attacks and paralyses the larva, lays an egg on its body and then seals the petrified larva into its burrow by blocking the entrance with a pebble.

Some of the true wasps practise a similar form of parasitism. They capture and paralyse spiders, burying them with their fertile eggs in a sealed cell. The solitary potter and mason wasps construct small cell-like receptacles from clay and saliva for this purpose. The cells are usually attached to plant stems and inside each one they deposit a paralysed caterpillar on which they lay their eggs before sealing the entrance. Sandwasps kill caterpillars to eat as well as providing them for their offspring. They bury the caterpillars in nesting holes dug in the ground, and after laying the egg on its crippled prey the sandwasp closes the hole, rams the sand into place with a pebble and then smoothes it over. This remarkable activity is made all the more unusual by the fact that the sandwasp manages to rediscover the nesting hole and repeats this performance at intervals to drop another caterpillar into her feeding young.

All the sandwasps and digger wasps construct some sort of cell which they stock with paralysed but living food for their larvae. The insects are either stung into immobility or their nerve centres are crushed. The mournful wasp, in funereal black, makes use of abandoned beetle holes and stocks its cell with aphids. Other species capture small flies or beetles and most of them dig their cells underground, some of them in extremely hard and compact soils.

Many of the bees also construct this type of underground burrow nest but only a few are parasitic. The cuckoo bees, sometimes called homeless bees, lay their eggs in the cells of other bees, relying on stealth to gain entry into the hive. When their grubs hatch they are fed by the host bees but later they also eat the other eggs and larvae. Some parasitic bees attack and kill young bumble bee queens, using the workers of the colony as slaves. Their behaviour is similar to the robber ants which capture pupae from the nests of other species. These ants are then born into slavery and spend their lives as workers to their captors.

The comparisons between this sort of behaviour and the evil aspects of some of our own history are both superficially too obvious and too perilous in cultural terms to warrant further discussion here. Nevertheless some of the macabre and bizarre situations to be found in the world of parasites defy even man's ingenious cunning. The concept of taking shelter in the body of a living animal is one which challenges our credulity, and in the complex life cycle of the liver flukes almost beggars description. The flukes inhabit the livers of many animals, feeding on their blood. They can inflict serious damage, resulting in death, and in this sense cannot be considered very highly evolved as parasites, but they undergo so many changes of form and habitat in their development that they must nonetheless have evolved during an ancient lineage. During their life cycle they require a number of hosts to complete the chain of their development. In some species they have to spend at least part of their life in a snail, followed by an amphibian, then a rat or mouse, before the adult emigrates to take up residence in a mammal such as a dog or a weasel. A less complex life style, but one which suggests the critical developments at these various stages, is that of the liver fluke *Fasciola hepatica* which infests sheep and cattle.

The worm-like adult flukes, about one centimetre long, have no fixed form and alter the shape of their body to suit the confines of their environment inside the sheep's liver. Here they deposit thousands of eggs in one batch, which are passed out with the sheep's

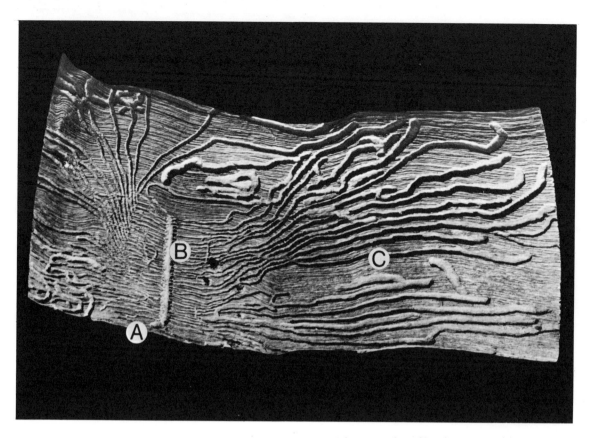

A large number of beetles are parasitic upon plants, and the works of the bark beetles create distinctive patterns upon many trees.
These carvings are caused by the ash bark-beetle, *Hylesinus fraxini*. A The small mating chamber, entered by the male and female through a hole in the bark. B The egg gallery leading from the mating chamber is carved out by the female. An egg is laid in each of the notches on either side. C Tunnels radiating from the egg gallery are carved out by the larvae, each avoiding its neighbour, and growing larger as they develop. The tunnels are cut partly in the wood and partly in the bark, so that both are grooved.

Left and below The solitary mud-dauber wasp builds long semi-tubular structures, each of which is subdivided into separate compartments containing an egg and which are stocked with paralysed spiders as future provisions for its young. The nest pipes are guarded by the male while the female collects building materials and hunts for prey.

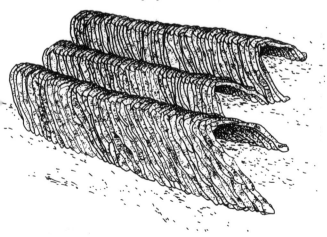

The larvae of each species of gall wasp makes a characteristic gall. The wasp lays her egg in the plant tissue and when the larva hatches it exudes a secretion which stimulates the plant to form a gall. The illustrations here are of British species, all, apart from the rose bedeguar, found on the oak.

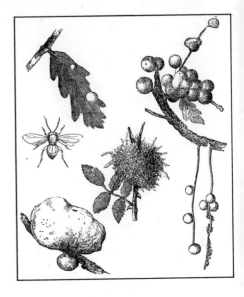

faeces on to the grass. From each egg some hundreds of young will be born, but hatching does not take place unless there is rain or heavy dew. In dry conditions however the eggs can survive for periods of up to a year. After hatching the young flukes disperse, swimming along the film of moisture on the grasses or migrating to ponds and streams. When they locate a particular species of snail they burrow into its skin and there undergo the second stage in their development. The juvenile fluke becomes blind, loses its tail, and changes into an egg-shaped sedentary animal. Soon afterwards it becomes transformed into a small crawling creature and makes its way to a new residence in the host snail's digestive gland. It then, remarkably, develops offspring of its own kind known as redia. Other equally strange developments occur within different fluke species. Some hatch from the egg as miniature forms of the adult, whereas in another species the larvae live only as pairs united in permanent copula. With the liver fluke however the redia once again change shape to resemble miniature versions of the adult. These gnaw their way out of the snail and swim off to attach themselves to a plant at the water's edge. Here they can live for up to six weeks, enclosed in a protective cyst, waiting to be eaten by a sheep. When they have been swallowed they hatch out inside the sheep's stomach and migrate to its liver, there to begin again the destructive cycle of self-survival.

CHAPTER NINE

The Snare of the Hunter

Over 80 per cent of all animals are insects. Nearly one million have been named but there are probably some three million species in existence. They are the natural food of many birds and other animals and billions of them are eaten daily. Their rapid rates of reproduction are not sufficient to ensure survival against these odds and the insects have therefore evolved many methods of avoiding attack.

One of the more effective of these methods is by camouflage or cryptic colouring. Examples of insects which closely resemble their natural surroundings are widespread and common. Some disguise themselves against lichens and tree bark, are shaped like thorns on a branch or take the form of shrivelled leaves. Others appear indistinguishable from bird droppings and many insect larvae resemble twigs in colour, form and posture. A great number of insects are green-coloured and are thus more easily hidden among the vegetation, and many other animals are coloured to resemble or blend with their surroundings. White animals for instance proliferate in the Arctic while desert animals are often sand-coloured. Animals with striped or dappled coloration live among forests and some fishes resemble seaweed or stones.

A much rarer method of concealment is by masking. This is best seen in the spider crabs which live among seaweed. They shred the weed like strips of paper, soften it and make it sticky by chewing, then rub it firmly on to their head and legs. If placed in a tank with no weeds the spider crabs place shingle on their backs. Similarly centipedes roll their sticky eggs in sand to disguise them with a covering of earth. The green caterpillar of the palisade sawfly feeds on poplar leaves and is able to protect itself by erecting a palisade around the leaf. This is formed of dried saliva resembling a mould growth. The palisade does not enable the caterpillar to resist attacks but succeeds in keeping out small predators by suggesting that the leaf is too unwholesome to investigate. Birds' eggs are commonly camouflaged by pigments deposited on the eggs as they pass through the oviduct and usually found in such birds as plovers and curlews which lay their eggs in exposed places. For this reason the eggs of the black-headed gull are well-camouflaged. The interior of the egg however is glaringly white; thus when a chick is hatched the broken pieces of eggshell are hastily removed from the nest area to avoid attracting predators.

The main function of solitary breeding is also for concealment, and in those species where the male is brightly coloured the nest-building and incubation is normally carried out by the dull-coloured and camouflaged female. Concealment is also achieved by disguising the nests. Grebes for instance build inconspicuous floating platforms and when they leave the nest cover their whitish eggs with pieces of vegetation, while female partridges use their cryptic body colouring to conceal the eggs. The vividly coloured scarlet tyrant of South America builds a nest covered with lichens and bound together with spiders' webs so that it blends perfectly with its surroundings.

The ultimate form of protection occurs when the eggs or young are carried about by the parent. The male midwife toad bears the ropes of spawn tangled round his thighs, and with Darwin's frog the male swallows the eggs, which develop in his vocal sacs. Some species of spiders and scorpions carry their young on their backs. Young kangaroos live in their parent's pouch and the pouch of the bandicoot faces to the rear so that the baby is protected from dirt and stones as the mother digs for food. The echidna carries her young in a pouch, but after their spiny covering begins to develop she lays them under bushes or in some hiding place while she forages for food, returning to suckle them.

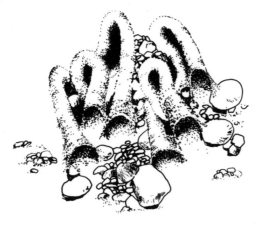 

Snares built by the larvae of *Hydropsyche* and *Neureclipsis* caddis flies. These larvae live in rapid streams and the traps are always constructed to face upstream.

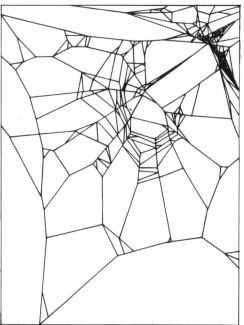

An orb-web built by a *Zygellia* spider. The gap in the spiral threads is characteristic of the genus. Compare the symmetry of this construction with the drawing (*above*) which shows a *Zygellia* web constructed in a laboratory after the spider had been fed with a dose of caffeine. Studies of the effects of drugs on web-building have been carried out at the North Carolina Department of Mental Health in a programme related to the sensory motor mechanisms of spiders.

The helpless young of many mammals, unlike newly hatched birds, seem generally to be born without any instinctive understanding of alarm or fear. When disturbed or exposed they will often simply move about in circles, whimpering perhaps but making no attempt to conceal themselves or to escape. Fortunately they are normally well-hidden and protected by their parents, but it is unusual to find parental care outside the birds and mammals. With a few notable exceptions young reptiles, amphibians, insects and fishes have to care for themselves as soon as they hatch. They are all well-adapted by some means or other for survival. Often they immediately seek some form of shelter, and with the larvae of the caddis flies this has evolved into the building of a protective casing around their bodies.

Almost all caddis fly larvae live underwater but one species, which is found only in Worcestershire, is completely terrestrial and lives round the base of large trees. The caddis fly larvae exhibit a great variety of construction techniques. Some build their tubes with sticks, others of stones, sand grains, shell fragments or whatever material is available. One Victorian naturalist, a Miss Smee, even managed to get them to build with gold dust. They also build cases of compacted sand grains around themselves. These cases are bound around the larvae with sticky secretions or strands of silk and some species also construct silken nets for catching their food.

Traps, snares and nets are commonly employed by many animals for catching others. Some of the solitary wasps, as we have seen, paralyse their victims and store them in a closed cell as live meat for their grubs when they hatch. The larvae of the tiger beetles dig traps where they sit and wait for passing insects, and the ant-lion larvae excavate pitfalls. These pitfalls are always constructed in dry sandy areas so that any insect inspecting the edge of the pit will dislodge the loose sand and tumble down into the exposed jaws of the larva. If the victim does not fall into reach the ant-lion larva hurls sand at it as it struggles to escape, knocking it off balance and into the base of the trap. When drained of nourishment the carcass is flung out of the hollow.

Nets and snares may seem particularly vicious methods of catching live prey, but they are only extensions of other terrifying weapons such as claws, teeth and beaks. In fact the spider's web becomes an essential extension of its tactile organs. The earliest spiders however were not web spinners. The ancestral spider probably used its silk threads as a drag line each time it ran out from its hiding place to seize passing insects. Eventually these lines would form a network of threads radiating from the lair and inevitably would serve to act as trip wires, thus forming the origin of the spider's web. This form of radiating network closely resembles the webs spun today by the most primitive of the spiders, found in China and Japan, but the simplest construction yet discovered, though not necessarily primitive, is nothing more than a single horizontal thread from about one to four metres in length with a central sticky section. As soon as an insect brushes against the thread the spider jerks forward, simultaneously releasing the tension and jolting the thread, thus entangling its prey.

Another remarkable method of capture is adopted by a spider which makes only a small web about the size of a postage stamp and which it holds limp in its four front legs while watching for its victim. When an insect comes near the spider suddenly stretches out the web to its widest span, hurls forward and wraps the net around its prey in gladiator fashion. There is another species which never spins a web but nonetheless makes effective use of its silk glands. It drops a short line of thread which has a small sticky globule of silk on the end. Holding this thread in one of its legs it then whirls it around like a helicopter blade for about 15 minutes, after which time the line is hauled in and eaten along with any insects which may have collided with and stuck to the globule. A few minutes later a new line is spun and the process repeated.

But the spider does more than catch food with its silken threads. It wraps and preserves this food in silk sheets; it hibernates in silk-lined chambers and it flies on silken threads. Indeed its ability to spin silk has helped the spider to become so successful that it is now distributed all over the world and the family has over 40,000 different representatives. A spider's silk glands are continually spinning and wherever the animal goes it lays down a silken thread. The thickness of this thread is determined by a control valve at the base of the spinneret, and the spider is also able to produce different types of thread for different purposes.

Diagrammatic elevation of an orb-web showing the various components.

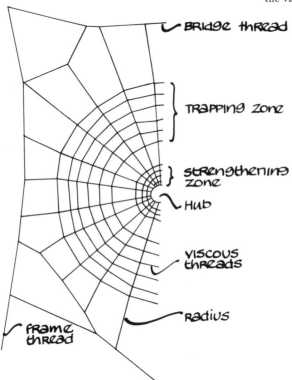

François Xavier Bon de Saint Hilaire, who had a pair of mittens and some stockings made from spider's silk, wrote in 1709 his *Dissertation on the Usefulness of Spider's Silk*. Following this publication Réaumur was invited by the Académie des Sciences to investigate and report on the silk-spinning properties of these animals. There were enthusiastic hopes that spider farms could be established for silk production, but Réaumur quashed these visions by pointing out the problems of having to keep the cannibalistic spiders apart from one another. Furthermore only the silk from the egg cocoons is suitable for spinning and Réaumur calculated that 27,648 female spiders would be required to produce one pound of silk. His report was later translated into Manchu at the request of the Emperor of China, partly so that he could satisfy his own curiosity but also to ensure that no serious rivals to the jealously guarded silkworm production were about to appear in Europe.

Because the spider's silk has no commercial value the spider's web received little scientific attention until John Blackwall in 1835 produced his paper *On the Manner in Which the Geometric Spiders Construct Their Webs*. But it was not until 1910 that Richard Hingston, experimenting with orb-web spiders, discovered that the integral design of the web is a consequence of the size of the spider, for there is a relationship between leg length and web size and between body weight and size of mesh. Young spiders with relatively long legs in proportion to their light body build large webs with narrow mesh. Older spiders are heavier but do not have a comparative increase in leg length and their webs are composed of larger meshes. In addition as the animal becomes heavier it has to build with thicker threads to take its weight. The thickness alone would not affect the web pattern but the increased amount of material that has to be produced for these threads means that shorter threads are used to construct the web. But these are not the only determining factors and despite many years of patient and close study it is amazing how little is still known of many aspects of the spider's web. Today slow motion films, series photography and mathematical-statistical analysis are being used in an attempt to determine exactly how the web is constructed; why different threads are placed in certain positions; and what determines or upsets the patterns of construction. Studies of the spider's central nervous system in relation to its web-building capabilities are being made, and in the USA and France investigations based on web structure are assessing the ability of the spider to learn and remember.

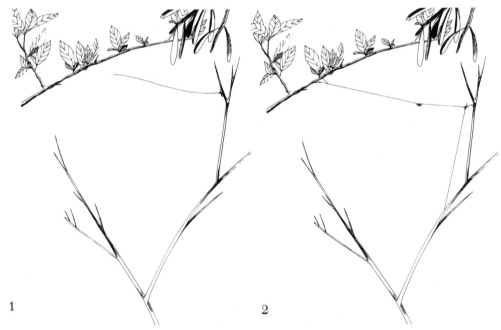

1

2

Above and opposite The stages of construction in building an orb-web. Using the wind to carry its thread (1) the spider first sets out the bridge line, which is then made secure as necessary before the spider begins to stretch another strand across the gap (2). Dropping from the centre of this strand it then anchors a vertical thread (3) forming the apex of the construction and establishing the point to which it will always return between each stage of the construction. The spider now quickly secures a series of radial lines (4) and when these are completed begins spinning the temporary dry thread spiral (5). Finally, using this auxiliary spiral as a guide line, it retraces its steps and places the permanent sticky thread spiral (6). While doing this the spider is rolling in the temporary spiral and eating it, its left hind leg is pulling new silk from the spinnerets, and its right front leg probes forward to the next radial line.

120

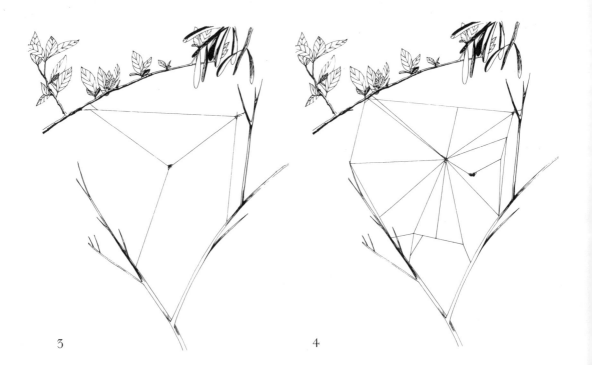

3   4

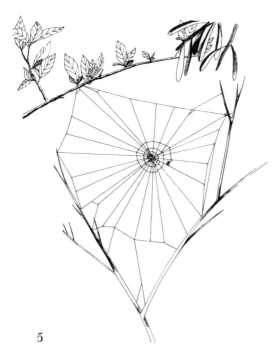 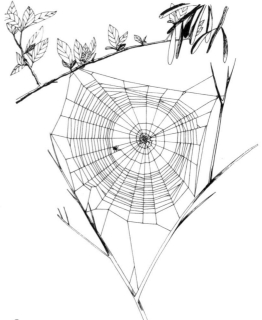

5   6

121

Opposite *Araneus diadematus* at the centre of its web waiting for prey to collide with its trap. The white dots are small fragments left over from the auxiliary spiral.

Lateral view of a sheet-web of *Cyrtophora citricola*; a tensile structure visually reminiscent of some of the latest developments in human technology.

The web is used as a method of measuring the spider's behaviour because the spinning of silk is such a total and continuous expression of its behaviour. One of the problems of investigating webs under construction is that this activity normally takes place in the small hours of the morning. Stimulant drugs were first used in an attempt to encourage activity at a more convenient hour of the day for research. The only result was that the spiders persisted in building webs in the early morning darkness but produced crazy and misshapen structures. Various hallucinatory-inducing substances such as LSD and mescalin have also been fed to spiders as part of a research programme investigating hallucination in mental diseases. The reactions and changes of behaviour were recorded by the change in the structure of the web, the spider in effect writing out its own results.

The results of these experiments by Dr Peter Witt are interesting and intriguing. With doses of amphetamine the spider produced a smaller web than normal and its pattern was grossly distorted. Mescalin also resulted in short amounts of thread being produced and the web size and regularity of angles were affected. The effects of caffeine were a lengthening in the perimeter, a significant increase in the number of over-sized central angles and a decrease in the catching area. On high doses of caffeine web-building was so disturbed that the results were barely recognisable as webs. Strychnine diminished the frequency of web-building, as did D-lysergic acid Diethylanide. But small doses of LSD also resulted in a general increase in angle and spiral regularity, an increase in the size of the catching area and a decrease in the number of over-sized angles; in short a more 'perfect' web than the normal one.

To attempt to explain what the nervous system of any animal is programmed to achieve requires an answer which is proportionally more elaborate in relation to the complexity of the animal's behaviour. But when considering relatively simple levels of behaviour the explanation becomes more direct. One can say for instance that the nervous system of a spider is programmed to build a web. This directness does not diminish the complexity of such an operation, it simply allows us to investigate this phenomenon rather more precisely. For the spider beginning the construction of a new web and suspended in space from a few framework threads, there is only a minimum of guidance. The threads it lays out are at one and the same time both its structure of intent and of information. The careful positioning of each line is a guide to each subsequent stage and the eventual shape of the web emerges as a series of responses. Beyond the existence of an instinctive programme for web-building the spider seems to have the ability to sample and evaluate the progress of its construction. Young spiders especially will often reposition threads, lengthening or shortening them to best suit the location, and they seem to construct trial structures as a preliminary to complete web-building.

The most elaborate and most common web is the orb web spun by the common garden spider. This web is spun within a framework of radiating threads attached to the boundaries and forming a central hub. Thus the radii are joined to elastic supports which allow the whole structure to flex in the wind or when struck by a flying insect. The various components of the web each have a specific role, the structural elements being the hub, the radii and the frame. These are constructed from dry threads which unlike the threads used in making the spiral are not sticky and do not coalesce when they touch. Therefore whenever the structural threads of the web intersect the spider has to join them together. This ensures a continuity of structure so that any disturbance in one part of the web is immediately transmitted along all the threads. Stop motion photography has shown that when the spider is trying to locate just where a fly is struggling in the web it pulls at the structural threads, setting up a series of elastic pulsations. It is thus able to locate the position of its prey by detecting the change in tension. It is also probable that this action serves to entangle the victim in the mesh as well as setting up a motion which would make retaliatory attack difficult.

The construction of a web begins when the spider lays out a few long lines almost, apparently, at random, for this is followed by a series of complex tests and adjustments as threads are tightened, expanded and repositioned. Sometimes they are entirely removed, the line being detached and then rolled up and eaten. During this process the line serves as a temporary support for the spider while it lays out a new thread, the spider's body acting almost as a zip between the two. Eventually a central hub is established and this acts as a stage point for all subsequent operations. Between stages of build-

ing the spider continually returns to the hub and tests the strength and state of the structure. Early in the course of laying out the radii threads the spider has completed the frame of the web and the radii lines are attached directly to this frame. Occasionally a short tie line is fixed near the end of a radius thread, taken along the frame and pulled taut so that a Y-shape is constructed, giving the radius in effect a double attachment. In between spinning each of the radii the spider spends some time resting at the centre of the web. It is likely that during this time more protein-rich silk is being pumped into the animal's silk ducts, for the building of a web temporarily empties its ampullate glands. While it is waiting the spider circles and grasps each of the radii threads in turn, seeking out oversized sectors with its front legs and testing the tensile state of the web. The pattern of the web is tested and completed entirely by touch; a spider building within a light-tight box is in no way handicapped by lack of visual stimulus.

When the spider has felt its way round a complete circuit of the hub without having detected the necessity for any further additions, the course of construction then changes from a radial to a spiral pattern of construction. A broadly spaced provisional spiral is laid out around the web. When this is completed it serves as a guide and support to the spider as it turns around and travels back on a spiralling journey to the hub. As it journeys it unravels the provisional spiral and lays down in its place a viscid thread which forms the catching area of the web. The animal uses its legs for both locomotion and thread guidance during the spiral spinning, pulling the sticky thread from its spinnerets with its back legs and stretching forward with its front legs to probe the

Far left Each species of spider spins a characteristically different web. This is the tangled silk snare built by *Linyphia montana*. Insects flying into the upper threads are knocked down into the thickly woven sheet where they are attacked and captured.

An engineering achievement and a thing of beauty: the central platform of a spider's web spangled with dew drops on a misty morning.

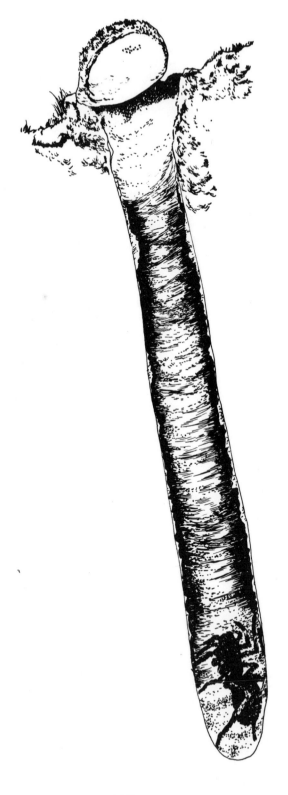

The trapdoor spider creates an ingenious dwelling by excavating a vertical shaft with its jaws and sealing it with a camouflaged, hinged lid (here shown opened), neatly bevelled to fit tightly in the neck of the tunnel. The tiny puncture marks on the inside of the door are made by the spider's fangs as it holds it tightly shut against intruders. The tunnel, drawn approximately full size, is coated with a waterproof lining of saliva-moistened earth and this is then covered with a dense sheet of spun silk. The tunnel provides seclusion from sun and rain as well as predators. It also serves as a courtship parlour, a nuptial chamber and a nursery. The spider rarely leaves its tunnel, and even when lunging outside to capture passing insects it usually keeps the door propped open with its back legs for a safe and rapid retreat.

The dramatically simple web of *Hyptiotes cavatus*. The spider holds the apex thread at the top right-hand corner and uses the web as a net rather than a trap. The structure has to be rebuilt after each capture.

next point of attachment. Finally when the spiral has been completed the spider makes a short course of inspection at the hub, removing any tangles and criss-crossing any holes until the hub is completed as a dense platform of small bridging threads on which it then sets up station to await the first early callers.

Although the web components and pattern of structure change with the age of the individual, spiders spin webs which are characteristic of their species, and the snares and webs of different spiders vary greatly in form and structure. These differences probably have little relation to the effectiveness of the different webs for catching food and have probably evolved as means of signalling species differentiation between the sexes. Considering the cannibalistic habits of the females such an early warning system seems essential to the survival of the species. As it is a male will take anything up to one hour in approaching his potential mate.

The orb web of the common garden spider is certainly the most well-known of all the webs and the most thoroughly investigated. But the webs of the house spider are perhaps even more familiar. They are usually built in corners and are of triangular shape, the apex formed as a silk tube in which the owner sits and waits for its visitors. Another species which is a close relation to the house spider lives out of doors, usually on gorse bushes, and spins a horizontal web above which it lays out a tangle of individual threads. These act as a baffle to flying insects, knocking them down into the sheet below where the spider sits. These webs are produced by a member of the group of spiders which includes the so-called money spiders and are the most abundant species. This type of construction sometimes covers an area twice the size of this page with the tangled superstructure above rising for about a foot. On the other hand some representatives of this family which live on the ground spin mere scraps of silken sheets between the grasses. On occasions, when certain weather conditions prevail, all such spiders in a meadow will simultaneously spin new webs. When this occurs a dramatic effect is created, with a continuous carpet of shimmering silver stretched over the landscape.

Spiders' webs can be classified into five main sections: first there are the *irregular* webs, such as the maze of threads built by the house spider; the *sheet* webs which are closely woven in a single plane with no regular arrangements; the *funnel* webs of which the principal part is also a sheet but also frequently includes an irregular net woven above; the *orb* webs, the characteristic feature of which is a series of radiating lines supporting a viscid spiral of silk; and finally there are *triangular* webs.

As mentioned previously, silk threads are also used for a variety of other purposes. Young spiders use their threads for kiting on the wind. The minute voyagers, like a speck of pepper seed on the air, migrate on silky filaments up to a metre in length. In Germany they are called *der fliegender Summer*, the flying summer. Not all spiders spin webs with their threads. The jumping crab spiders for example hide and leap out upon their prey, as does the wolf spider. But they also often spin a thread on which they can haul themselves back to their lair with their catch. The large and hairy spiders of the tropics often attack anything they encounter and some are extremely vicious. One only has to pass within a few steps of the hide and they leap out to give chase. W. H. Hudson told how he was almost caught one day while out riding on the pampas: 'I suddenly observed a spider pursuing me, leaping swiftly along and keeping up with my beast. I aimed a blow with my whip, and the point of the lash struck the ground close to it, when it instantly leaped upon and ran up the lash, and was actually within three or four inches of my hand when I flung the whip from me.'

All spiders hunt and live in solitude and seem to be ever hungry and ever watchful.

They are extraordinarily successful but at first glance would not seem to be a very safe bet in the evolutionary stakes. Their natural weapons of defence are often not particularly effective; they are not very strong for their size and their body is unprotected and fleshy. But whereas man achieved success as a hunter because he learned to use primitive tools against other creatures of greater speed and strength, so the spiders have also developed artificial aids to trapping. And perhaps none is more curious than the trapdoor spiders. They are mainly nocturnal animals which dig deep tunnels for themselves, just wide enough to allow the spider to turn around inside, and are lined with silk. The entrance to this tunnel is shut with a closely fitting lid. Two types are constructed: one is a thin circular sheet of silk, hinged as a flap; and the other is like a cork, much thicker and with a bevelled edge, fitting like a plug into the tube. To make the silk door the spider covers the entrance with a closely woven layer of silk, then bites away the edges leaving one part as the hinge. The cork type of trapdoor also begins as a covering of silk over the entrance. The spider then carries earth and lays this on top, binding this down with more silk and continuing the process until the plug is thick enough. Finally it camouflages the entrance with bits of moss and vegetable matter. The spider hides safely concealed in its burrow rarely leaving this shelter except to dash out and capture passing insects.

The different forms of nests built by spiders are even more numerous than their webs. Some make nests in the earth, varying from a simple vertical shaft with a delicate lining of silk to one in which the entrance is built up to resemble a small turret. Many species make nests by rolling leaves into a tube and lining the space with silk or use the leaves as a foundation for spinning a silk chamber around them to protect the eggs. Even more remarkable are the silken, tubular nests of the purse-web spider found in the south-west USA. These spiders live in tunnels in the ground, invariably at the base of a tree. The tunnel, about as deep as a man's hand, is lined with silk which then extends as a vertical tube above the surface for about half a metre. The outer lining of the tube is coated with minute particles of lichen and bark to camouflage it with the tree against which it is built. The top of the tube is flattened where it is attached to the bark by bands of threads. Between the points of contact with the tree and the ground the tube is stretched taut. Thus any insect walking upon the structure causes it to vibrate whereupon the spider rushes up the tunnel, bites through the web and captures the insect, pulling it inside. The damaged tube is later repaired as necessary and the spider returns to wait in its hole in the ground.

Spiders certainly deserve our observation and admiration. Though some degree of affection towards them might not be possible, many of them offer special interest to us and sometimes a degree of charm. Invisible aeronauts flying in the wind and sunshine on filmy threads, or the English garden spider like a gem-stone in its dew-spangled web, certainly appeal more to our aesthetic sense than does the leaping wolf spider or the heavy bird-eaters on their hairy stilted legs. But acts of chilling cruelty can be carried out by what are to us more visually attractive species, including, unfortunately, man himself.

CHAPTER TEN

Design and Decoration

Among the many strange and wonderful activities of birds perhaps the most enchanting to human eyes is the manner in which some species decorate their nests. Starlings often gather sycamore blossoms and sprays of forget-me-nots; sparrows add white alyssum flowers to their untidy nest structures, and tits place fragments of torn leaves round the nest cup. Even such an unlikely bird as the jackass penguin collects an odd variety of objects with which to decorate its burrow. Sea shells and pebbles are the usual choices, but feather quills, pieces of rope and driftwood are also used.

The significance of these actions is not yet fully understood and it is particularly mystifying to discover these actions in such an unexpected place as an ants' nest. This remarkable yet apparently useless activity was first investigated by P. I. Marikovsky, the Russian entomologist, after he had noticed that wood ants often carry bright and shiny objects into their nest. Normally these would be such objects as beetle wings, which obviously have no food value. Marikovsky therefore tipped a handful of small coloured beads on top of the nest to see the ants' reaction. Immediately there semed to be great excitement. Some of the ants, as one would expect, began to clear the beads away as they would any other rubbish, but other individuals hurriedly carried the beads inside the nest, occasionally having to grapple with another who intended to clear it out of the vicinity. Different reactions were observed at different nests, apparently depending on the age of the community. On one occasion 2,000 beads were promptly removed into the nest within four minutes, whereas at other nests the ants almost totally ignored the beads, which were only removed as they were discovered along with other debris which had accumulated on the nest.

Marikovsky came to the conclusion that the difference in behaviour between the nests was due to the fact that new colonies were too busy with the more important tasks of building and tending the brood to be bothered with such trivia as glass beads.

Old-established communities on the other hand would contain a relatively large number of individuals who were not particularly busy and were more readily stimulated by something new and different. It might be that the beads could be considered more as some kind of toy rather than an item of decoration, for some of the beads are carried out of the nest each day and dumped by the more serious and diligent workers.

Occasionally nest decoration has more obvious and strictly functional purposes. The crested bell-bird of Australia drapes large hairy caterpillars around its nest. The caterpillars are paralysed by squeezing and when they are dead and withered the bell bird replaces them with freshly plundered bodies. They are not, fortunately, used for mere decoration but as a deterrent to any predator, for many caterpillars can spray irritant fluids or taste unpleasant. But the most macabre decorations are the larders of the shrikes. These birds have earned their common name of 'butcher birds' from their habit of impaling their victims on thorn trees or barbed wire. Their most common prey are large insects but they also catch frogs, lizards, rodents and other birds. Hanging their prey on thorns may have been originally for the purpose of a hook to hold the victim while it was being eaten. However like many predators the shrikes kill more than they need and their larders become overstocked and rotting. In parts of Germany the shrike is known as the 'ninekiller' because it is said it kills nine singing birds each day.

Of the remaining examples of nest decoration the results are more attractive and the reasons more pleasurable. As with the arts of primitive man, decoration and ceremony have become integrated items of expression for some species of birds. And though their 'art forms' may be purely instinctive they are nonetheless beautiful to us and, as we shall see, no less attractive to the female birds.

Ritualistic nest-building is something common to many animals as a displacement activity. When two individuals face each other in argument on the border of their territories, their rage, instead of being directed against each other, sparks off a series of actions which precludes the possibility of inflicting damage but acts as a safety valve to their pent up energy and frustration. Embarrassed or irritated human beings pull and stroke their hair in an indecisive or unsettling situation, just as starlings if confronted with a rival will preen their feathers. With sticklebacks the gestures are more extreme but no less controllable. Two males, warning each other at the boundary of their territories, will reach such a pitch of fury that they both suddenly dive into a vertical position and then, standing on their heads, start to dig holes in the gravel in the way in which they dig their nests. The viciousness of their digging in such a situation is in proportion to the fury of their disturbance. This applies equally to herring gulls which instead of digging holes attack the grass around them and gouge chunks from the ground in place of a murderous attack on their neighbour.

These ritualised activities are essential mechanisms for survival, and in some species of birds the rituals have become integrated performances of their territorial behaviour. This is clearly shown with the 'arena birds', some 85 species which include sandpipers, grouse, bustards, manakins and birds of paradise. To a greater or lesser extent each of these species displays arena behaviour, which is the establishment, by a group of males, of a mating territory made up of a number of individual display areas. Within their courts the males then perform displays of mock fighting, sometimes called 'Lek displays', from the Swedish *leka* which means 'to play' but with sexual connotations. The male birds establish their courts by singing, dancing and playing with sticks. In this way they avoid fighting each other, unlike 99 per cent of the world's birds.

Arena behaviour however is not confined to birds. It can be seen in animals as widely separated as the Uganda kob and the cicada-killer wasp, but it is in birds that it is expressed most clearly and most often. Perhaps the most exotic ambassadors for the arena birds are the South American cocks-of-the-rock. These jaunty vivid birds assemble for exuberant displays of dancing on the jungle floor when the males attempt to attract responses from the females. The arena is usually an area about 15 by 30 metres and is defended by all the males, about 40 in number, each occupying a small private court among the bushes. Each male clears his court of debris and dead leaves by violently flapping his wings and thus creates a stage for his exhibition. Spreading wide his scarlet-orange wings and tail he steps out and begins to move in a gentle swaying

Unlike human architects, animal builders use materials only for their functional qualities. They are not concerned with high fashion or styles, but their artificial structures are beautiful examples of structural integrity. The changes in use of material at different stages of construction for different purposes are clearly shown in this nest built by a buff-throated saltator, a South American member of the finch family. The outer shell of the nest is composed of dry twigs with strips of broad-shaped leaves for the middle layer and finer material for the inner lining. The birds shape the nest with their bodies.

Nest of Gray's thrush, the Central American equivalent of the British song thrush, built on a palm leaf. The bulky nest built by this bird has an inner layer of mud and in the rainy months of early summer fragments of living vegetation mixed with this mud take root and send out green shoots, transforming the nest structure into an attractive aerial garden.

motion. The rhythm and excitement of his dance increases until it becomes a combination of astonishing leaps and gyrations. Finally he retires, whereupon another male begins his performance, and so on.

Arena displays, being carried out by so many widely differing species all over the world are a clear example of convergent behaviour and in almost all the species concerned the males have evolved some form of exaggerated adornment for the display. The Old World ruffs for example take their name from their extravagant collar or ruff of feathers around their necks. In the spring the male ruffs gather in groups around small grass-covered hillocks. Each male occupies a small territory about one metre in diameter,

and here they display their adornment of feathers in quaint almost courtly gestures whenever the female reeve appears. She then wanders around them and all the males stay motionless, ruffs spread out, beaks stooped to the ground. Finally she chooses a suitable male, signifying her acceptance by pecking him. Immediately he springs to life and copulation takes place, whereupon all the other rejected males fall flat on their faces and lie as if mortally wounded.

Similar arena behaviour, but with the courtly dances presented on a larger scale, is shown by the prairie chicken, a bird which at one time was widespread across the United States. Their drastic reduction has been caused not so much by deliberate destruction of the birds as by destruction of their ancestral display arenas or 'booming grounds'. A prairie chicken arena can be almost one kilometre in length by 200 metres wide and include 400 males who space themselves some 5–15 metres apart. Starting in early spring the males begin their dawn and dusk dances, stamping their feet noisily and rapidly, pivoting, flashing tail feathers and eyebrows and all the while giving a steady booming note. The best position on the parade ground for giving a display is obviously along the centre line. In one study 74 per cent of the visiting females chose to give their favours to the four males in this location. Only after this stalwart quartet were satiated did the females then begin to take notice of the adjacent males, while those dancing on the periphery of the arena were totally ignored.

There are many other instances of birds assembling for dancing and singing. Sometimes the participants appear most unlikely candidates for such behaviour. The black-faced ibis of Patagonia for example, a bird almost as large as a turkey, dashes and flings itself about in a late evening caper, screaming wildly as it does so. Rails also indulge in wild screeching dances, rushing about with widespread wings, but the dance of the American jacanas is a more practised and leisurely exhibition. Their dance floors are smooth levelled areas surrounded by vegetation. They have dazzling wing feathers, greenish-gold, and they hold these high like grouped flags of silk as they move gently around each other on their long thin toes.

Superb displays are also given by the Australian lyrebirds. The males defend large territories and within this area create a number of display courts. Damp soil is raked into shallow mounds on the forest floor, approximately one metre in diameter, and upon these each lyrebird sings and dances, vibrating his long and delicate tail feathers in a shimmering cascade of silver and mauve and prancing to the rhythm of his own melody. The most spectacular and colourful displays of all however are those of the exotic birds of paradise of which there are 40 species inhabiting New Guinea and northern Australia. The males perform elaborate and exuberant displays to attract the females, some high in the trees, often hanging upside down or even holding themselves out horizontally from a branch. Others clear spaces on the forest floor for their dances and spend many hours clipping foliage from the bushes to allow shafts of sunlight to fall on to their display courts and illuminate their iridescent feathers.

A great deal of attention is given to these dance courts by the birds, clearing away leaves and stones from the stage. It is but a little step away from such actions to *adding* leaves and stones to the stage as striking decoration against the cleared earth, and the impulses of court clearing and court decorating are probably deeply rooted in common ceremonies.

The bower birds are closely related to the birds of paradise and share the same habitats. Like them they also have courts for displaying and there the males attract and copulate with as many females as possible. Unlike the birds of paradise however the

Left The best zoological gardens aim to provide facilities which encourage the animals to carry out their patterns of natural behaviour. This construction was built by a satin bower-bird in the 'World of Birds' at the Bronx Zoo.

bower birds, although brightly coloured, have not evolved such extravagant and fabulous plumage. Instead they build theatres on their courts and decorate them with trinkets. There is a definite correlation between the elaboration of the birds' plumage and the complexity of the bowers constructed. The most simple bowers are built by birds with decorative crests, while the rather dull brown bower bird, or crestless gardener, of which the male is identical in appearance with the female, builds a most ornate construction with an overhanging roof and collects heaps of shells, berries and flowers to decorate his court.

The development of the bower bird's creations may be influenced by play as well as ritualistic display. True play activities are widespread among many birds. There are convincing accounts of such activities among species as widely dissimilar as hornbills, buzzards, road runners and parrots. Kestrels often play with pine cones, and warblers play with a variety of objects. These birds also engage in tilting contests, just as gyrfalcons seem to enjoy mock aerial combat. Eiderducks play at shooting the rapids and rooks are often seen soaring on currents of air, achieving no purpose but apparently indulging in the activity for the pure delight of the sensation.

Many of the games which are carried out by water birds may well be only a part of some complex display activity, and some of the elaborate displays enacted by the bower birds while in their bowers probably have some sort of function. Equally however these activities may be displays of a playful nature, for there is evidence which suggests that these 'displays' have progressed beyond mere functionalism and are carried out, at least in part, for pleasure.

The male bower birds in each of the species always have some sort of stage, and carpet this with decorative objects. They also always build within audible range of one another and spend a great deal of time whistling and calling, not only to attract females but also to keep in touch, for they all seem to know when there is a female in the vicinity. The males, as with all arena birds, are chained to their display grounds for many months of the year and take no part in nest-building or care of the young. The female builds the cup-shaped nest of twigs on her own; whereas nest-building activities in the male have survived in the form of the bower.

There are three distinct types of bower-building behaviour: stage-building; avenue or wall-building; and maypole-building. The stagemakers construct the simplest type of bower in relation to some of the other species. Typically they are cleared areas of ground on the jungle floor, one to two metres in diameter, on which a few colourful objects and up to about 40 fresh leaves are placed. The leaves are deliberately placed upside down to make them more conspicuous and withered leaves are cleared from the stage each day.

The avenue or wall builders are the most widespread group, occurring throughout New Guinea and parts of Australia. In its simplest form an avenue bower is a narrow open corridor of sticks built on a carefully laid mat of vertical sticks. At the entrance to this corridor the male collects a colourful selection of feathers, pebbles, shells and fruits. The colours of the objects are often chosen with discrimination but the shapes seem immaterial. If opportunity offers the birds will add aluminium camping spoons, shards of broken glass, silver coins and, on one occasion, a farmer's glass eye. The wealth of material collected by each male is illustrated by the contents found in one bower built by a Lauterbach's bower bird. This included over 1,000 pale-coloured pebbles, 3,000 sticks and 1,000 thin wisps of grass. In this species the sticks are interlocked to form not merely an avenue but a rigid four-walled structure and the grass is used to line the

To humans it appears that this tern's nest site has been decorated with a collection of sea shells. The effect of the bird's design, however, is to camouflage its solitary exposed egg.

The superb glossy starlings of East Africa usually build their nests in thorny acacia trees, and add further protection to their nest site by decorating the tree with extra quantities of thorn twigs placed around the nest and on branches leading to it.

Bottom   A spotted bower bird waits at the entrance to its bower for a prospective mate. To attract admirers the bird has collected shards of green glass to decorate the front porch of his distinctive construction.

internal walls. Inside this court the male displays to the female, strutting about holding high a bright red berry.

An equally colourful display of objects is collected by the satin bower bird which constructs its bower in the Australian forests. The ground in front of the walled bower is decorated with any object the bird can find which is either green or blue, to blend with its plumage. Lobelia blossoms, delphinium petals, cornflowers, parrot feathers, broken crockery, blue beads and scraps of paper are all added to the display. Very often these objects are stolen from neighbours and equally often the bower bird quickly tears and breaks down his rival's bower before snatching a beakful of blue trophies and flying away.

The satin bower bird indulges in other odd aspects of behaviour which like vandalism and theft seem more human than animal. This bird is also a painter: using a wad of bark as a brush it daubs black paint on the walls of its bower. The paint is made of charcoal dust mixed with saliva and because this does not have very lasting qualities the satin bower bird is constantly having to renew its decorations. Bower-painting, which is probably a displacement activity for courtship feeding, also occurs with the regent bower bird and the spotted bower bird. Each of these species mixes crushed vegetable material and saliva to make a green paint which it smears on to the interior walls with its beak. Another unusual feature of the satin bower bird's bower is that it is always orientated to the sun on a north-south axis, thereby catching the bright early morning sun when its displays are at their most energetic.

Most accomplished of all the bower builders however are the maypole builders. The drably coloured and appropriately named brown bower bird belongs to this group. But the wonderful bower which it builds puts even the most wildly decorated bird of paradise's display to shame. At the base of a sapling on a cleared piece of flat ground the brown bower bird builds a cone of moss about the size of a man's fist. He then constructs a radiating pile of twigs around the sapling, leaving an overhanging aperture to serve as entrance. Other twigs are thrust into the structure to make it firm and eventually the conical structure is so thickly thatched that it is completely waterproof. A garden of moss is then carefully laid out in front of this hut; flowers and brightly coloured fruits, all carefully selected for colour, are arranged on the lawn.

All the maypole builders construct their bowers on the ground around the trunk of a sapling, some building huts with a garden fence around. The golden bower bird builds a tower up to three metres high. As with the brown bower bird a display court of moss is laid out around the base of the tower and decorated with small colourful objects, such as snail shells, beetles' wing cases, and fresh flower heads which have to be changed daily for many months.

With abilities for decorating and building which mirror our own sense of aesthetics so closely, it is understandable that Dr Odoardo Beccari, the first naturalist to discover the display court of a bower bird, should have believed that he had found a playhouse built by native children. Nor does it seem surprising that one of the nineteenth century naturalists should have suggested that all the birds should be divided into two categories; bower birds and other birds.

However, the creations of the bower birds, no less than the building works of any other animal, are stimulated by purely functional requirements. Furthermore no matter how apparently insignificant any of them may seem to our eyes, each animal structure is a masterpiece in its own right, and although we may dismiss many of their attempts as crude or simple they nonetheless play some role in the development of our concept of architecture.

We should remember that the perpetual forces of evolution have determined the wide diversification of building activities which each species has developed to ensure its survival, and those examples which we have examined here – from the fortified constructions of the ovenbirds to the underwater nests of the sticklebacks, from the towering structures built by termites to the gossamer skeleton of the spider's web – all serve to remind us that, in every example of successful architecture, form is dictated by function.

Sources of Reference

AGRAWAL, V. C, 'Field Observations on the Biology and Ecology of the Desert Gerbil *Meriones hurrianae* in Western India' in *J. Zool. Soc. India*, Vol 17, 125–34, 1965.

AKIMUSHKIN, I. I, *Cephalopods of the Seas of the USSR*, Jerusalem, 1965.

ARDREY, R, *The Territorial Imperative*, New York, London, 1966.

ARISTOTLE, *History of Animals*, tr Richard Cresswell, London, 1862.

ARMSTRONG, A. E, *Bird Display*, Cambridge, 1942.

AUSTIN, O. L. Jr, *Birds of the World*, New York, 1961; London, 1965.

BARNES, V. G. Jr and BRAY, O. E, 'Black Bears use Drainage Culverts for Winter Dens' in *J. of Mammology*, Vol 47, Kansas, 1966.

BATES, H.W, *The Naturalist on the River Amazon*, London, 1864; reprinted Gloucester, Massachussetts, n.d.

BEEBE, C. W, *The Bird*, New York, 1906.

BELL, T, *A History of British Quadrupeds*, London, 1874.

BERGAMINI, D, *Land and Wildlife of Australia*, Life Nature Library, 1965.

BLUNT, W, *The Compleat Naturalist*, London, New York, 1971.

BOLWIG, N, 'A Study of the nests built by Mountain Gorillas and Chimpanzees' in *South Africa J. of Science*, Vol 55, 1959.

BORRADAILE, L. A, *The Animal and its Environment*, London, 1923.

BROOKE, H, 'Britain's Habitats' in *Animals*, London, May–Sept 1969.

BURRELL, H, *The Platypus*, Sydney, 1927.

BURTON, M, *Animal Partnerships*, London, 1969; New York, 1970.

BUXTON, J, *The Oxford Book of Insects*, Oxford, 1968.

BUXTON, P. A, *Animal Life in Deserts*, London, 1923.

CAMERON, R, *Shells*, London, 1961.

CARR, D. E, *The Sexes*, Garden City, New York, 1970; London, 1971.

CAULLERY, M, *Parasitism and Symbiosis*, London, 1952.

CHAUVIN, R, *The World of an Insect*, New York, London, 1967.

COLLIAS, N. E. and COLLIAS, E. C, 'An Experimental Study of the Mechanisms of Nest Building in a Weaverbird' in *The Auk*, No 79, 568–595, October 1962.

COLLIAS, N. E. and COLLIAS, E. C, *Evolution of Nest Building in the Weaverbirds (Ploceidae)*, Univ of California, Los Angeles, 1964.

COLLINS, W. B, *Empires in Anarchy*, London, 1967.

COMSTOCK, J. H, *The Spider Book*, New York, 1940.

DARWIN, C, *On Humus and the Earthworm*, London, 1945.

DAVEY, N, *A History of Building Materials*, London, 1961; New York, 1971.

DINES, A. M, *Honeybees from Close Up*, London, New York, 1968.

FARB, P, *Land and Wildlife of North America*, Life Nature Library, 1966.

FRASER, D, *Village Planning in the Primitive World*, London, n.d., New York, 1968.

GODFREY, G. and CROWCROFT, P, *The Life of the Mole*, London, 1960.

GRAAF, G. DE and NEL, J. A. J, 'On the tunnel system of Brant's Karroo rat *Paratomys brantsi* in the Kalahari Gemsbok National Park' in *Koedoe*, No 7, Pretoria, 1964.

HARRISSON, B, 'A Study of Orang-utan behaviour in the semi-wild state' in *International Zoo Yearbook*, Vol III, Zoological Society of London, 1961.

HEINROTH, O. and HEINROTH, K, *The Birds*, Univ of Michigan, 1959.

HICKSON, S. J, *An Introduction to the Study of Recent Corals*, Univ of Manchester, 1924.

HINDE, R. A, 'Nest building behaviour of Canaries' in *Proc. Zool. Soc. London*, No 131, 1–48, 1958.

HINDE, R. A, *Animal Behaviour*, New York, 1966.

HOWSE, P. E, *Termites: a study in social behaviour*, London, New York, 1970.

HUDSON, W. H, *The Naturalist in La Plata*, London, 1892; reprinted New York, 1968.

HUDSON, W. H, *The Book of a Naturalist*, London, 1919; reprinted New York, 1968.

KENDEIGH, S. C, *Animal Ecology*, Englewood Cliffs, New Jersey, 1961.

KEVAN, D. K. MCE, *Soil Animals*, London, 1962.

KIRMIZ, J. P, *Adaptations to Desert Environment*, New York, London, 1962.

KLASING, O, *Das Buch der Sammlungen*, Leipzig, 1875.

LACK, P, *Ecological Adaptations for Breeding in Birds*, London, 1968.

LANE, F. W, *Kingdom of the Octopus*, London, 1957; New York, 1965.

LEDOUX, A, 'Arboreal Nests' in *Proc. Xth Int. Cong. Entomology*, Vol 2, 521–528, Montreal, 1958.

LUBBOCK, J, *Ants, Bees and Wasps*, London, 1882.

MACDONALD, M. and LOKE, C, *Birds in the Sun*, Chester Springs, Pennsylvania, London, 1962.

MARTIN, H. T, *Castorolgia*, Montreal, 1892.

MELLANBY, K, *The Mole*, London, 1971; New York, 1972.

MILLAIS, J. G, *The Mammals of Great Britain and Ireland*, London, 1904.

MILLS, E. A, *In Beaver World*, Boston, 1913.

MOGGRIDGE, J. T, *Harvesting Ants and Trap-door Spiders*, London, 1873 (and Supplement to, 1874).

MORTON, J. E, *Molluscs*, London, 1967; New York, 1968.

NEL, J. A. J, 'Burrow Systems of *Desmodillus auricularis* in the Kalahari Gemsbok National Park' in *Koedoe*, No 9, Pretoria, 1966.

NICHOLS, D. and COOKE, J, *The Oxford Book of Invertebrates*, Oxford, 1971.

OMMANNEY, F. D, *Collecting Sea Shells*, London, 1968.

PRICE, N, *Winged Builders*, London, 1959.

RÉAUMUR, R. E, *The Natural History of Ants*, New York, 1926.

RUDOFSKY, B, *Architecture without Architects*, New York, 1965.

SAVORY, T. H, *The Spider's Web*, London, New York, 1952.

SAVORY, T. H, *Spiders, Men and Scorpions*, London, 1961.

SCHAPERA, I, *The Khoisan Peoples of South Africa*, London, 1930.

SCHMIDT-NIELSEN, K, *Desert Animals*, Oxford, 1964.

SEAGER, H. W, *Natural History in Shakespeare's Time*, London, 1896.

SKUTCH, A. F, *Life Histories of Central American Birds*, Berkeley, Calif. *Vol I* 1954, *Vol II* 1960, *Vol III* 1969.

SKUTCH, A. F, 'Nunbirds' in *Animal Kingdom*, Vol LXXII, No 5, New York, October 1969.

SKUTCH, A. F, 'Jacamars' in *Animal Kingdom*, Vol LXXIII, No 2, New York, April 1970.

SMEATHMAN, H, *Of the Termites in Africa and Other Hot Climates*, Phil. Trans. Royal Soc. Vol XV, London, 1809.

SMYTHIES, B. E, *The Birds of Borneo*, Edinburgh, 1960.

STEMPIER, M. F. Jr, 'Sponges' in *Animal Kingdom*, Vol LXXIII, No 4, New York, August 1970.

SUDD, J. H, *An Introduction to the Behaviour of Ants*, London, New York, 1967.

THOMPSON, D'ARCY, *On Growth and Form*, Cambridge, 1942.

THOMSON, A. L. (Ed), *A New Dictionary of Birds*, London, 1964.

THORPE, W. H, *Learning and Instinct in Animals*, Cambridge, Massachussetts, London, 1963.

TILETSON, J. V. and LECHLEITNER, R. R, 'Some comparisons of the black-tailed and white-tailed prairie dogs in North-Central Colorado' in *The American Midland Naturalist*, Vol 75, Univ of Notre Dame, Indiana, 1966.

TILQUIN, A, *La Toile Géométrique*, Paris, 1942.

TINBERGEN, N, *Curious Naturalists*, Oxford, 1958; Garden City, New York, 1968.

TINBERGEN, N, *Animal Behaviour*, Life Nature Library, 1965.

TURNER, N, 'Insects in Houses' in *Connecticut Agricultural Experimental Station Circular*, No 224 (revised), June 1970.

VAN SOMEREN, V. G. L, *Days with Birds*, Fieldiana: Zoology, Vol 38, Chicago, 1956.

WARBURTON, C, 'Arachnida embolobranchiata' in *The Cambridge Natural History*, Vol IV, London, 1923.

WARREN, E. R, 'The Beaver', monograph of the *Amer. Soc. of Mammalogists*, No 2, Baltimore, 1927.

WELLS, M. J, *Brain and Behaviour in Cephalopods*, Stanford, 1962.

WHITE, T. H, *The Book of Beasts*, London, 1954; New York, 1970.

WITT, P. N, REED, C. F. and PEAKALL, D. B, *A Spider's Web*, Heidelberg, 1968.

WOOD, J. G, *Homes without Hands*, London, 1866.

WOODBURN, J, *Hunters and Gatherers*, British Museum, 1970.

WORRELL, E, *The Great Barrier Reef*, London, New York, 1967.

Index

Acorn barnacle, 21
African lungfish, 70
Alligators, 57; nest of, 57
Amoebocytes, 80
Animals: aquatic, burrowing as shelter, 63;
behavioural patterns of, 14, 16, 18;
burrowing by, **70**; nervous system
achievements, 123; partnerships in, 103;
reaction of to surroundings, 14, 16;
stimulus of, 14, 16
Ant-chat, 12; nest of, 12
Ant-eaters, 12
Ants, 23, 81: army, 87; behaviour of, 81;
blackjet, 86; carpenter, 86; castes of, 82;
common red, 105; elbowed red, 105;
harvester, 86; honey-collecting by, 105;
honeypot, 86; *Iridomyrmex humilus*, 107;
industriousness of, 85; *Leptothorax
emersoni*, 105; *Myrmica canadensis*, 105;
nest of, 82, 85; parasitism among, 105;
parasol, 85; pharaoh's, 107; robber, 112;
Solenopsis, 105; tailor, 86–7, nest of, 86–7;
weaving (*Oecophylla smaragdina*), **84**;
wood, 33, 107, 129, decoration of nests by,
129; nest of, **83**
Archaeopteryx, 41
Arctic hare, 12
'Arena birds', displays and exhibitions by,
130–3
Argonaut, **21**, 22
Aristotle, claims regarding pointed eggs, 25
Armadillos, 22, **28**, 39
Artificial external factors, effect of on
society, 81

Barber fishes, 103
Bats, 14
Bears, 12: dens of, 14; polar, cubs of, **15**
Beaver skins, export of, 78
Beavers, 14, 73–8: activities of, **76**, **77**;
building of dams by, 75; colonies, 75;
fame of, 75; lodge of, **76**, 78
Beccari, Dr Odoardo, 136; discovers display
court of bower-bird, 136
Bee-eaters, 29, 96
Bees, 14, 23, 57, 96: carpenter, 39; cell
construction of hive, 97, 99; comb, **100**,
cell angles in, 100–1, **102**; cuckoo, 111;
hive functions of, 96–7; leaf-cutter, 39;
mining, 39, 74; origin of, 96; potter, 39;

predators on, 96; solitary (*Osmia rufa*), nest
of, **38**, 39, 57, species of, 39
Beetles, 39, 40: bark, **113**; death-watch, 40;
dung, egg-depositing by, 23, 31; egg-laying
sites of, 23; elm-bark, 40; flour, 40;
furniture, 40; scavenging burying, 31, 33;
tiger larvae, 111, 118
Bell, Thomas, 38; description of mole's
fortress, 38
Bird's egg, purity of form of, 25
Bird's nest, popular conception of, 43
Birds: adoption of human habitat shelters,
14; decoration of nests by, 129–30; methods
adopted for egg-protection, 23–4; nesting of
in human habitations, 108; territorial
behaviour performances by, 130
Birds of paradise, 130, 133; displays by, 133
Bison, 71; near-extinction of, 71
Bitterlings, 68
Blackbirds, 108
Black-cowled oriole, nest of, **15**
Black-faced ibis, 133; dance behaviour by,
133
Black-headed gull, 116
Black kite, 109
Black-necked weaver, **45**
Blackwall, John, 120; Works, *On The
Manner in Which the Geometric Spiders
Construct Their Webs*, 120q.
Blue-footed booby birds, 41
Bon de Saint Hilaire, François Xavier, 119;
Works, *Dissertation on the Usefulness of
Spider's Silk*, 119q.
Bower-bird, 133–4: dull-brown, 134, 136;
golden, 136, nest of, 136; Lauterbach's, 134;
ritualistic display of, 134; satin, **133**, 136;
display of objects by, 136; odd behaviour of,
136; spotted, **135**
Bower-building behaviour, 134; types of,
134; avenue or wall, 134; maypole, 134, 136;
stage, 134
Brittle stars, 65
Brunel, Sir Marc Isambard, 64; studies
tunnelling methods of ship worms, 64
Brush turkeys, 46
Buff-throated saltator, **131**
Bühle, theory regarding shape of egg, 25–6
Bullfrogs, 57
Burrows, use of made by divers birds and
animals, 109